Randomized Response

STATISTICS: Textbooks and Monographs

A Series Edited by

Vol. 1: The Generalized Jackknife Statistic, *H. L. Gray and W. R. Schucany*
Vol. 2: Multivariate Analysis, *Anant M. Kshirsagar*
Vol. 3: Statistics and Society, *Walter T. Federer*
Vol. 4: Multivariate Analysis: A Selected and Abstracted Bibliography, 1957-1972, *Kocherlakota Subrahmaniam and Kathleen Subrahmaniam* (out of print)
Vol. 5: Design of Experiments: A Realistic Approach, *Virgil L. Anderson and Robert A. McLean*
Vol. 6: Statistical and Mathematical Aspects of Pollution Problems, *John W. Pratt*
Vol. 7: Introduction to Probability and Statistics (in two parts), Part I: Probability; Part II: Statistics, *Narayan C. Giri*
Vol. 8: Statistical Theory of the Analysis of Experimental Designs, *J. Ogawa*
Vol. 9: Statistical Techniques in Simulation (in two parts), *Jack P. C. Kleijnen*
Vol. 10: Data Quality Control and Editing, *Joseph I. Naus* (out of print)
Vol. 11: Cost of Living Index Numbers: Practice, Precision, and Theory, *Kali S. Banerjee*
Vol. 12: Weighing Designs: For Chemistry, Medicine, Economics, Operations Research, Statistics, *Kali S. Banerjee*
Vol. 13: The Search for Oil: Some Statistical Methods and Techniques, *edited by D. B. Owen*
Vol. 14: Sample Size Choice: Charts for Experiments with Linear Models, *Robert E. Odeh and Martin Fox*
Vol. 15: Statistical Methods for Engineers and Scientists, *Robert M. Bethea, Benjamin S. Duran, and Thomas L. Boullion*
Vol. 16: Statistical Quality Control Methods, *Irving W. Burr*
Vol. 17: On the History of Statistics and Probability, *edited by D. B. Owen*
Vol. 18: Econometrics, *Peter Schmidt*
Vol. 19: Sufficient Statistics: Selected Contributions, *Vasant S. Huzurbazar (edited by Anant M. Kshirsagar)*
Vol. 20: Handbook of Statistical Distributions, *Jagdish K. Patel, C. H. Kapadia, and D. B. Owen*
Vol. 21: Case Studies in Sample Design, *A. C. Rosander*
Vol. 22: Pocket Book of Statistical Tables, *compiled by R. E. Odeh, D. B. Owen, Z. W. Birnbaum, and L. Fisher*
Vol. 23: The Information in Contingency Tables, *D. V. Gokhale and Solomon Kullback*

Vol. 24: Statistical Analysis of Reliability and Life-Testing Models: Theory and Methods, *Lee J. Bain*
Vol. 25: Elementary Statistical Quality Control, *Irving W. Burr*
Vol. 26: An Introduction to Probability and Statistics Using BASIC, *Richard A. Groeneveld*
Vol. 27: Basic Applied Statistics, *B. L. Raktoe and J. J. Hubert*
Vol. 28: A Primer in Probability, *Kathleen Subrahmaniam*
Vol. 29: Random Processes: A First Look, *R. Syski*
Vol. 30: Regression Methods: A Tool for Data Analysis, *Rudolf J. Freund and Paul D. Minton*
Vol. 31: Randomization Tests, *Eugene S. Edgington*
Vol. 32: Tables for Normal Tolerance Limits, Sampling Plans, and Screening, *Robert E. Odeh and D. B. Owen*
Vol. 33: Statistical Computing, *William J. Kennedy, Jr. and James E. Gentle*
Vol. 34: Regression Analysis and Its Application: A Data-Oriented Approach, *Richard F. Gunst and Robert L. Mason*
Vol. 35: Scientific Strategies to Save Your Life, *I. D. J. Bross*
Vol. 36: Statistics in the Pharmaceutical Industry, *edited by C. Ralph Buncher and Jia-Yeong Tsay*
Vol. 37: Sampling from a Finite Population, *J. Hajek*
Vol. 38: Statistical Modeling Techniques, *S. S. Shapiro*
Vol. 39: Statistical Theory and Inference in Research, *T. A. Bancroft and C.-P. Han*
Vol. 40: Handbook of the Normal Distribution, *Jagdish K. Patel and Campbell B. Read*
Vol. 41: Recent Advances in Regression Methods, *Hrishikesh D. Vinod and Aman Ullah*
Vol. 42: Acceptance Sampling in Quality Control, *Edward G. Schilling*
Vol. 43: The Randomized Clinical Trial and Therapeutic Decisions, *edited by Niels Tygstrup, John M. Lachin, and Erik Juhl*
Vol. 44: Regression Analysis of Survival Data in Cancer Chemotherapy, *Walter H. Carter, Jr., Galen L. Wampler, and Donald M. Stablein*
Vol. 45: A Course in Linear Models, *Anant M. Kshirsagar*
Vol. 46: Clinical Trials: Issues and Approaches, *edited by Stanley H. Shapiro and Thomas H. Louis*
Vol. 47: Statistical Analysis of DNA Sequence Data, *edited by B. S. Weir*
Vol. 48: Nonlinear Regression Modeling: A Unified Practical Approach, *David A. Ratkowsky*
Vol. 49: Attribute Sampling Plans, Tables of Tests and Confidence Limits for Proportions, *Robert E. Odeh and D. B. Owen*
Vol. 50: Experimental Design, Statistical Models, and Genetic Statistics, *edited by Klaus Hinkelmann*
Vol. 51: Statistical Methods for Cancer Studies, *edited by Richard G. Cornell*
Vol. 52: Practical Statistical Sampling for Auditors, *Arthur J. Wilburn*
Vol. 53: Statistical Signal Processing, *edited by Edward J. Wegman and James G. Smith*
Vol. 54: Self-Organizing Methods in Modeling: GMDH Type Algorithms, *edited by Stanley J. Farlow*
Vol. 55: Applied Factorial and Fractional Designs, *Robert A. McLean and Virgil L. Anderson*
Vol. 56: Design of Experiments: Ranking and Selection, *edited by Thomas J. Santner and Ajit C. Tamhane*
Vol. 57: Statistical Methods for Engineers and Scientists. Second Edition, Revised and Expanded, *Robert M. Bethea, Benjamin S. Duran, and Thomas L. Boullion*
Vol. 58: Ensemble Modeling: Inference from Small-Scale Properties to Large-Scale Systems, *Alan E. Gelfand and Crayton C. Walker*
Vol. 59: Computer Modeling for Business and Industry, *Bruce L. Bowerman and Richard T. O'Connell*
Vol. 60: Bayesian Analysis of Linear Models, *Lyle D. Broemeling*

Vol. 61: Methodological Issues for Health Care Surveys, *Brenda Cox and Steven Cohen*

Vol. 62: Applied Regression Analysis and Experimental Design, *Richard J. Brook and Gregory C. Arnold*

Vol. 63: Statpal: A Statistical Package for Microcomputers – PC-DOS Version for the IBM PC and Compatibles, *Bruce J. Chalmer and David G. Whitmore*

Vol. 64: Statpal: A Statistical Package for Microcomputers – Apple Version for the II, II+, and IIe, *David G. Whitmore and Bruce J. Chalmer*

Vol. 65: Nonparametric Statistical Inference, Second Edition, Revised and Expanded, *Jean Dickinson Gibbons*

Vol. 66: Design and Analysis of Experiments, *Roger G. Petersen*

Vol. 67: Statistical Methods for Pharmaceutical Research Planning, *Sten W. Bergman and John C. Gittins*

Vol. 68: Goodness-of-Fit Techniques, *edited by Ralph B. D'Agostino and Michael A. Stephens*

Vol. 69: Statistical Methods in Discrimination Litigation, *edited by D. H. Kaye and Mikel Aickin*

Vol. 70: Truncated and Censored Samples from Normal Populations, *Helmut Schneider*

Vol. 71: Robust Inference, *M. L. Tiku, W. Y. Tan, and N. Balakrishnan*

Vol. 72: Statistical Image Processing and Graphics, *edited by Edward J. Wegman and Douglas J. DePriest*

Vol. 73: Assignment Methods in Combinatorial Data Analysis, *Lawrence J. Hubert*

Vol. 74: Econometrics and Structural Change, *Lyle D. Broemeling and Hiroki Tsurumi*

Vol. 75: Multivariate Interpretation of Clinical Laboratory Data, *Adelin Albert and Eugene K. Harris*

Vol. 76: Statistical Tools for Simulation Practitioners, *Jack P. C. Kleijnen*

Vol. 77: Randomization Tests, Second Edition, *Eugene S. Edgington*

Vol. 78: A Folio of Distributions: A Collection of Theoretical Quantile-Quantile Plots, *Edward B. Fowlkes*

Vol. 79: Applied Categorical Data Analysis, *Daniel H. Freeman, Jr.*

Vol. 80: Seemingly Unrelated Regression Equations Models : Estimation and Inference, *Virendra K. Srivastava and David E. A. Giles*

Vol. 81: Response Surfaces: Designs and Analyses, *Andre I. Khuri and John A. Cornell*

Vol. 82: Nonlinear Parameter Estimation: An Integrated System in BASIC, *John C. Nash and Mary Walker-Smith*

Vol. 83: Cancer Modeling, *edited by James R. Thompson and Barry W. Brown*

Vol. 84: Mixture Models: Inference and Applications to Clustering, *Geoffrey J. McLachlan and Kaye E. Basford*

Vol. 85: Randomized Response: Theory and Techniques, *Arijit Chaudhuri and Rahul Mukerjee*

Vol. 86: Biopharmaceutical Statistics for Drug Development, *edited by Karl E. Peace*

Vol. 87: Parts per Million Values for Estimating Quality Levels, *Robert E. Odeh and Donald B. Owen*

Vol. 88: Lognormal Distributions: Theory and Applications, *edited by Edwin L. Crow and Kunio Shimizu*

ADDITIONAL VOLUMES IN PREPARATION

Randomized Response
Theory and Techniques

Arijit Chaudhuri
Rahul Mukerjee

INDIAN STATISTICAL INSTITUTE
CALCUTTA, INDIA

MARCEL DEKKER, INC. New York and Basel

ISBN 0-8247-7785-9

MARCEL DEKKER, INC.
270 Madison Avenue, New York, New York 10016

Current printing (last digit):
10 9 8 7 6 5 4 3 2 1

PRINTED IN THE UNITED STATES OF AMERICA

To Bulu
—A. C.

To my parents
—R. M.

Foreword

This impressive monograph by Professors Arijit Chaudhuri and Rahul Mukerjee (of the Indian Statistical Institute, Calcutta) is indeed an authentic treatise on randomized response methods, covering the basic theory and diverse techniques under a common shade. Surveys involving sensitive queries, and the protection of privacy, on the one hand, and extraction of reliable information, on the other hand, laid down the foundation for randomized response methodology. The past twenty years have witnessed a phenomenal growth in the literature on randomized response methods. In fact, this has been one of the few areas in statistics where the theory and fruitful applications have emerged harmoniously, creating the need for a monograph emphasizing both theory and techniques. The current monograph is undoubtedly a useful milestone.

The authors are to be especially commended for their lucid presentation as well as their broad coverage of the subject matter. I have no doubt that this will be a valuable guidebook for people who would like to adopt randomized response methods in actual practice (and therefore need to familiarize themselves with the basic theory) as well as for researchers in statistics who would like to develop further methodology with a view to future applications. This is indeed a timely and scholarly undertaking by the two authors, and I have full appreciation for a job nicely done.

Pranab Kumar Sen
University of North Carolina
Chapel Hill, North Carolina

Preface

In socioeconomic investigations we sometimes need facts about highly personal matters which people usually like to hide from others. Also, an inquirer often feels a delicacy in asking direct questions about private and confidential subjects, especially if the subjects carry any social stigma. For example, he or she would feel uncomfortable asking a person about gambling, drug taking, tax evasion, or the extent of any illegal income, accumulated assets, history of induced abortion, and many similar items. Attempted open queries about such sensitive issues often result in high nonresponse rates and/or high response biases arising out of willful misstatements or blatant lies. So intelligent devices are needed to reduce rates of nonresponse and biased response in order that fruitful inferences may be drawn from survey data when the issues involved demand protection of privacy.

A simple technique involving the use of a randomized rather than a direct response was introduced by Warner (1965) and popularized by several other researchers who followed his lead. It was a modest beginning, visualizing a population dichotomized according to a sensitive characteristic, with the intention of estimating the proportion of persons bearing this characteristic. Warner's scheme solicits a randomized response of "yes" or "no" to a query about bearing the characteristic with a probability, and about its negation with the complementary probability. The respondent is expected to feel it to be safe to cooperate and is assumed to tell the truth. From a sample of such randomized responses (RRS) an estimate easily follows. This pioneering work led to modifications and developments in various directions, and there ensued around this powerful idea a tremendous spurt in research activity. Soon, polychotomous populations claimed a coverage allowing diverse alternative modes of reply. To cover the quantitative variables in addition to the categorical ones was, of course, a simple step. An important issue to settle was the question of striking a balance between the requirements of efficient estimation and protection of privacy by the correct choice of parameters specific to a randomizing device.

Estimation in RR started by employing the method of moments. Application of the maximum likelihood principle required some caution, leading to certain modifications. Other sophisticated principles of estimation, using linear models, Bayesian arguments, Hoeffding's U statistics, Von Mises' differentiable statistical functions, and so on, are gradually being applied.

In most activities involving RR, only simple random sampling is permitted. Moreover, an infinite population setup is usually postulated. But emerging ramifications cover selection as well, with varying probabilities from truly finite populations with identifiable units. Mathematical niceties thus far developed around RR procedures appear inherently stimulating to theoreticians. But the real worth of any RR technique lies in its practicability. Most devices proposed so far are reported to have been tried out successfully in real-life problems.

The copious growth of literature around the theory and techniques of RR within a decade of its inception called for an evaluation of the potentialities of the subject in a session convened by the International Statistical Institute in 1975 in Warsaw. The resulting proceedings, published in an issue of Volume 44 of the *International Statistical Review*, influenced further progress at an accelerated pace. An important survey paper by Horvitz *et al.*(1975) provided further impetus. After the elapse of a decade with uninterrupted activities in this area, in this monograph we review the history of RR. A concise and abridged version of some aspects of this is to appear as a review paper by Chaudhuri and Mukerjee (1987).

The present book is organized essentially in the following manner. We start with the simplest problem and its solution together with its properties, and then proceed to cover further developments, stating results plus their proofs when the latter are not too difficult. Certain examples are worked out, followed by exercises for the readers. When this is done, the relevant topics are meant for survey practitioners and undergraduate students covering elementary courses in sample surveys and theories of probability and inference. More complicated themes are meant for graduate students and researchers. Exercises are not given on topics at these levels. To avoid a mix-up of simpler topics with relatively complicated refinements thereof, we present the former in a chapter, leaving the latter to an appendix that immediately follows it. References are included for both chapter and appendix text.

We believe that a growing popular interest justifies this publication. It is gratifying to note that Campbell and Joiner (1973) recommend the inclusion of RR theory and methods in introductory undergraduate level courses in statistics and probability. If this suggestion is heeded, we believe an emerging generation of survey practitioners may gain some benefits from a text of the present type. For such a course, if offered, our Chapters 1, 2, and 3, Sections 5.1 to 5.3 of Chapter 5, and Chapter 8 may merit consideration. The remainder of the text is intended for readers with higher-level commitments and aspirations. If necessary, one may consult relevant portions of the texts by, for example, Hoel *et al.* (1971), Cox and Hinkley (1974), Rao (1973), Kullback (1959), Serfling (1980), and Raghavarao (1971).

Our labor will be deemed amply rewarded if it attracts an extensive readership that shares our enjoyment of the subject, and stimulates further popular curiosity about RR. Comments on our lapses and suggestions for improved coverage will be gratefully appreciated.

Arijit Chaudhuri
Rahul Mukerjee

REFERENCES

The abbreviation RR (randomized response) and certain other standard abbreviations are used in the references throughout.

Campbell, C., and Joiner, B. L. (1973). How to get answers without being sure you've asked the question? *Amer. Statist.* **27**, 229–232.

Chaudhuri, A., and Mukerjee, R. (1987). RR techniques: a review. *Statist. Neerlandica* **41** (in press).

Cox, D. R., and Hinkley, D. V. (1974). *Theoretical Statistics.* Chapman & Hall, London.

Hoel, P. G., Port, S. C., and Stone, C. J. (1971). *Introduction to Probability Theory.* Houghton Mifflin, Boston.

Horvitz, D. G., Greenberg, B. G., and Abernathy, J. R. (1975). Recent developments in RR designs. In: *A Survey of Statistical Design and Linear Models*, ed. J. N. Srivastava. North-Holland, Amsterdam, pp. 271–285.

Kullback, S. (1959). *Information Theory and Statistics.* John Wiley, New York.

Raghavarao, D. (1971). *Constructions and Combinatorial Problems in Design of Experiments.* John Wiley, New York.

Rao, C. R. (1973). *Linear Statistical Inference and Its Applications*, 2nd ed. John Wiley, New York.

Serfling, R. J. (1980). *Approximation Theorems of Mathematical Statistics.* John Wiley, New York.

Warner, S. L. (1965). RR: a survey technique for eliminating evasive answer bias. *J. Amer. Statist. Assoc.* **60**, 63–69.

Acknowledgments

The present text is an expanded version of a preliminary draft which was well received by an anonymous reviewer, to whom we are grateful for constructive suggestions.

We owe a deep debt of gratitude to Professor P. K. Sen, University of North Carolina, Chapel Hill, whose active cooperation and interest greatly encouraged us to go ahead with our work on this book. We also greatly appreciate the encouragement of Professor J. K. Ghosh of our Institute.

We are also thankful to the professors in charge of our respective divisions, Professors T. J. Rao and S. Bandyopadhyay, and to Professor A. C. Mukhopadhyay, Head, Computer Science Unit, who liberally extended to us all the facilities that we needed for our work.

The helpful cooperation of Dr. Arun K. Adhikary of our Institute in preparing the manuscript is also recognized.

Finally, it is our pleasure to acknowledge the care with which the typing was done by the late D. K. Bardhan, Mr. P. K. Sen, and Mr. K. K. Saha.

Arijit Chaudhuri
Rahul Mukerjee

Contents

Randomized Response

1

Introduction to Randomized Response: The Warner Model

1.1. INTRODUCTION: WHY RANDOMIZED RESPONSE?

Socioeconomic investigations often relate to certain personal features that people wish to hide from others. In comprehensive inquiries, detailed questionnaires include numerous items. Data on most of them are frequently easy to procure merely by asking. But a few others may be on sensitive issues for which people are not inclined to state honest responses. For example, most people prefer to conceal the truth regarding their savings, the extent of their accumulated wealth, their history of intentional tax evasion and other illegal and/or unethical practices leading to earnings from clandestine sources, crimes, trade in contraband goods, susceptibility to intoxication, expenditures on addictions of various forms, homosexuality, and similar issues which are customarily disapproved of by society.

Open or direct queries often fail to yield reliable data on such confidential aspects of human life. Nonresponse or false or evasive responses to attempted direct queries about such private matters are so pronounced in practice that it is difficult to measure and control their effects to make an effective use of the acquired data from samples to reach a correct and fair conclusion about populations. So instead of open surveys, alternative procedures are needed if we are to procure reliable data on such confidential matters, especially the sensitive ones, believed to carry reprehensible stigmas. Randomized response (RR) survey techniques introduced by Warner (1965) provide such an alternative to meeting the twin objectives of generating enough reliable data to yield fruitful inference and creating a feeling among respondents that their privacy is protected despite their truthful replies to cleverly designed questions which do not reveal individual identities in the course of the survey.

1.2. THE WARNER MODEL

Consider the simple setup of a dichotomous population. Every person in the population belongs either to a sensitive group A, or to its complement, $\bar{\mathrm{A}}$. The problem is to estimate π_A $(0<\pi_A<1)$, the unknown proportion of population members in group A. To do so, a simple random sample with replacement (SRSWR) of size n is drawn from the population. Because of the sensitive, possibly stigmatizing nature of the characteristic under study, a direct question regarding membership in A or otherwise is not expected to be helpful in terms of cooperation from the respondents. Hence the results, based on such a direct question, are quite likely, to an inestimable extent, to be vitiated by bias due to untruthful and evasive answers, or more important, there may be excessive refusals, resulting in too small a sample size, thus reducing the level of efficiency of an estimator based on such survey data.

To overcome the difficulty noted above, Warner (1965) suggests the technique of randomized response (RR). In this procedure, each respondent is provided with a randomization device by which he or she chooses one of the two questions "Do you belong to A?" or "Do you belong to $\bar{\mathrm{A}}$?" with respective probabilities P and $1-P$ $(0<P<1)$ and then replies "yes" or "no" to the question chosen. In applications, of course, equivalent and appropriate vocabulary is to be used for the questions. The process of selecting one of the two questions is unobserved by the interviewer. Also, the interviewee is not to disclose the question to which his or her answer corresponds. Thus although the interviewer gets a "yes" or "no" reply, because of the randomization procedure, he or she cannot identify a particular respondent with group A or $\bar{\mathrm{A}}$ on the basis of such a reply. This protects the privacy of an interviewee, and thus he or she may now be expected to cooperate and respond truthfully. It may be noted that the probability P is chosen by the interviewer as a part of the design. Further details regarding an appropriate choice of P are considered later in this section.

As a device for randomization, each interviewee may be provided with an identical spinner fitted with a pointer. By drawing angles at the center, the face of the spinner may be so depicted that the pointer indicates the question relating to membership of A with probability P and the question relating to membership of $\bar{\mathrm{A}}$ with probability $1-P$. This is illustrated in Figure 1.1. It is then enough to ask each respondent to spin the spinner and

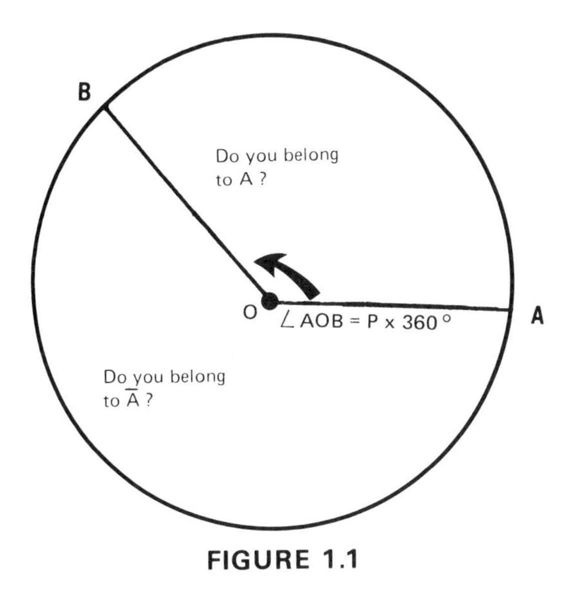

FIGURE 1.1

reply "yes" or "no" truthfully to the question indicated by the pointer. Infact, one can imagine numerous other devices for randomization. For example, instead of using a spinner, each respondent may be supplied with a deck of cards bearing the question regarding membership in A or $\bar{A}$ in proportions P and $1-P$ respectively. He or she may then be asked to draw one card after shuffling the deck thoroughly and to reply to the question inscribed on the card drawn.

The data generated through RR may be utilized in estimating π_A as follows. Assuming truthful reporting (in fact, throughout this book truthful reporting will be assumed unless otherwise stated), clearly the probability of getting a "yes" response is, say,

$$\begin{aligned}\lambda &= \pi_A P + (1-\pi_A)(1-P)\\ &= (1-P) + (2P-1)\pi_A \qquad (1.1)\end{aligned}$$

Here it is assumed that an outcome of the randomization experiment implemented through an appropriate device does not depend on the respondent's characteristics. Denoting the number of "yes" responses in the sample by n_1, an unbiased estimator (UE) of λ is $\hat{\lambda} = n_1/n$, the sample proportion of "yes" responses. Hence, by (1.1), a UE of π_A follows as (of course, taking $P \neq 1/2$)

$$\hat{\pi}_{AW} = \frac{P-1}{2P-1} + \frac{n_1}{(2P-1)n} \tag{1.2}$$

The choice of P is at the sampler's disposal. Clearly, $P=1$ amounts to an open or direct response (DR) survey, in which case $\hat{\pi}_{AD} = n_1/n$ is a UE of π_A with a variance

$$\text{var}(\hat{\pi}_{AD}) = \frac{\pi_A(1-\pi_A)}{n}$$

In the randomized case, observing that n_1 follows binomial distribution with parameters n and λ, it follows from (1.2) that

$$\begin{aligned}\text{var}(\hat{\pi}_{AW}) &= \frac{\lambda(1-\lambda)}{n(2P-1)^2} \\ &= \frac{\pi_A(1-\pi_A)}{n} + \frac{1}{n}\left[\frac{1}{16(P-0.5)^2} - \frac{1}{4}\right]\end{aligned} \tag{1.3}$$

Note that the expression for $\text{var}(\hat{\pi}_{AW})$ involves the unknown parameter λ. Therefore, to form an idea about the sampling fluctuation of $\hat{\pi}_{AW}$ from the sample itself, one has to develop an estimator of the variance above. Since

$$E(n_1) = n\lambda, \qquad E(n_1^2) = n\lambda + n(n-1)\lambda^2$$

where E stands for the operator of expectation, one gets

$$E\left[\frac{\hat{\lambda}(1-\hat{\lambda})}{n-1}\right] = E\left[\frac{nn_1 - n_1^2}{n^2(n-1)}\right] = \frac{\lambda(1-\lambda)}{n}$$

Hence a UE of $\text{var}(\hat{\pi}_{AW})$ is obtained as

$$\begin{aligned}\hat{\text{v}}\text{ar}(\hat{\pi}_{AW}) &= \frac{\hat{\lambda}(1-\hat{\lambda})}{(n-1)(2P-1)^2} \\ &= \frac{\hat{\pi}_{AW}(1-\hat{\pi}_{AW})}{n-1} + \frac{1}{n-1}\left[\frac{1}{16(P-0.5)^2} - \frac{1}{4}\right]\end{aligned} \tag{1.4}$$

For a DR survey, $\hat{\text{v}}\text{ar}(\hat{\pi}_{AD}) = \hat{\pi}_{AD}(1-\hat{\pi}_{AD})/(n-1)$ is a UE for $\text{var}(\hat{\pi}_{AD})$.

Observe that (1.3), treated as a function of P, is minimum if $P=0$ or 1, and the resulting variance naturally equals $\text{var}(\hat{\pi}_{AD})$. With P chosen close to 0 (or 1), $\hat{\pi}_{AW}$ will have a high efficiency, as the second term in (1.3) will be small, but a “no” (or “yes”) response will imply with a high probability that the interviewee then belongs to A and thus there will be little protection of his or her privacy and hence little possibility of truthful reporting. On the other hand, with P close to 1/2, the respondent is well protected but $\text{var}(\hat{\pi}_{AW})$ blows up. In particular, if $P=1/2$, the interviewee gets the maximum protection but π_A becomes nonestimable, since, by (1.1), λ no longer involves π_A. To summarize, efficiency and maintenance of confidentiality are in conflict in the context of RR inquiries. This point is elucidated further in Chapter 5.

Example 1.1. Warner’s technique of RR was applied in a recent survey to study the prevalence of alcoholism among undergraduate college students in Calcutta. A student was defined to be an alcoholic if he or she had consumed at least 1250 ml of alcoholic liquid over the 1-month period preceding the interview. With this definition, the objective was to estimate π_A, the proportion of alcoholics, on the basis of an SRSWR of size 100. It was decided to take $P=0.25$. Accordingly, each interviewee was provided with a deck of 60 cards, 15 of which bore the question “Did you consume at least 1250 milliliters of alcoholic liquids during the last 1-month period?”, while the remaining 45 cards carried the question “Did you consume less than 1250 milliliters of alcoholic liquids during the last 1-month period?”. Each interviewee was required to draw one card after shuffling the deck, unobserved by the interviewer, and then to reply “yes” or “no” truthfully to the question on the card drawn. There were altogether 72 “yes” and 28 “no” responses.

In this case we have $n=100$, $n_1=72$, and $P=0.25$. So $\hat{\lambda}=n_1/n=0.72$ and by (1.2), the Warner estimate of π_A is $\hat{\pi}_{AW}=0.06$. From (1.4), an unbiased estimate of $\text{var}(\hat{\pi}_{AW})$ is obtained as 0.008145.

EXERCISES

1.1. Verify the formulas (1.3) and (1.4) in detail.

1.2. Show that (1.3) can be expressed as

$$\text{var}(\hat{\pi}_{AW})=\frac{\pi_A(1-\pi_A)}{n}+\frac{P(1-P)}{n(2P-1)^2}$$

1.3. (Bourke and Dalenius, 1974) In the Warner model, suppose that truthful responses come not from all possessing A but only from a proportion $H(P)$ of them, namely the honest ones. $H(P)$ is termed the "honesty function." Assuming that those in $\bar{A}$ are all truthful, show that the probability of a "yes" response is

$$\lambda = \pi_A H(P)(2P-1) + (1-P)$$

Hence if $\hat{\lambda}$ is the sample proportion of "yes" responses, conclude that a UE of π_A may be obtained as

$$\hat{\pi}_A = \frac{\hat{\lambda} - (1-P)}{H(P)(2P-1)}$$

provided that $H(P)$ is known and $P \neq 1/2$.

1.4. In the setup of Exercise 1.3, proceed along the line of (1.3) and (1.4) to obtain the variance of $\hat{\pi}_A$ and an unbiased estimator of this variance.

REFERENCES

Bourke, P. D., and Dalenius, T. (1974). RR models with lying. *Tech. Rep.* 71, Institute of Statistics, University of Stockholm, Sweden.

Warner, S. L. (1965). RR: a survey technique for eliminating evasive answer bias. *J. Amer. Statist. Assoc.* **60**, 63–69.

Appendix

Supplementary Remarks on the Warner Model

A1.1 RANDOMIZED RESPONSE VERSUS DIRECT RESPONSE

Randomized response techniques (RRTs) were introduced by Warner (1965) mainly to cut down the possibility of (1) reduced response rate, and (2) inflated response bias experienced in direct or open surveys relating to sensitive issues. Warner himself pointed out how one may get a biased estimate in an open survey when a population consists of individuals bearing a stigmatizing character A or its complement $\bar{A}$, which may or may not also be stigmatizing.

Let an SRSWR of size n be taken and each sampled person be advised to respond 1 or 0 to imply, respectively, that he or she bears A or $\bar{A}$. As an estimator for π_A, the proportion bearing A, one may take, assuming that everybody responds

$$\hat{\pi} = \sum_1^n \frac{Y_i}{n}$$

where $Y_i = 1$ (0) if the ith respondent responds 1 (0). Denoting by T_a and T_b the respective probabilities that a person truthfully states that he or she bears A or $\bar{A}$, one has

$$E(\hat{\pi}) = \pi_A T_a + (1 - \pi_A)(1 - T_b)$$

Then the bias of $\hat{\pi}$ is

$$B(\hat{\pi}) = \pi_A(T_a + T_b - 2) + (1 - T_b)$$

and the variance of $\hat{\pi}$ is

$$\text{var}(\hat{\pi}) = \frac{\{\pi_A T_a + (1 - \pi_A)(1 - T_b)\}\{(1 - \pi_A T_a) - (1 - \pi_A)(1 - T_b)\}}{n}$$

However, it is unlikely that one will have any way to guess correctly about the magnitudes of T_a and T_b so as to be able to judge the extent of bias involved and the effect on the accuracy in estimation. If, instead, an RRT is employed in the situation above as explained in Chapter 1, one has a way of estimating π_A unbiasedly and getting an unbiased estimator as well for the variance of that estimator as given by Warner (1965).

A1.2 UNBIASED ESTIMATION IN THE WARNER MODEL

Incidentally, we may note an objection from Moors (1985) regarding the unbiasedness of Warner's estimator. Referring to relation (1.1) in Chapter 1, we may note that since $0 \le \pi_A \le 1$, one has

$$1-P \le \lambda \le P \quad \text{if} \quad P > \tfrac{1}{2} \qquad \text{or} \qquad P \le \lambda \le 1-P \quad \text{if} \quad P < \tfrac{1}{2}$$

Although it is a probability, λ is not permitted values outside these intervals, depending on the value of P. In other words, the entire interval [0,1] is not a relevant parametric space for λ, which thus has a truncated parametric space.

Moors (1985) demands that a parameter estimator take no values outside the convex hull of the parametric space. The Warner estimator $\hat{\pi}_{AW}$ for π_A obviously does not meet this constraint. He also shows that there exists no UE, under this restriction, for π_A related to λ through (1.1), except when $P=0$ or 1. As an amendment to $\hat{\pi}_{AW}$ valued outside [0, 1], Raghavarao (1978) proposes a shrinkage-type estimator and examines its bias and efficiency.

A1.3 MAXIMUM LIKELIHOOD ESTIMATION WITH THE WARNER MODEL

Using the notation of Section 1.2, the likelihood for λ given the sample of "yes-no" replies is

$$L = \binom{n}{n_1} \lambda^{n_1} (1-\lambda)^{n-n_1} = L(\lambda)$$

and the likelihood for π_A is

$$L \propto [(1-P)+(2P-1)\pi_A]^{n_1}[P-(2P-1)\pi_A]^{n-n_1} = L(\pi_A)$$

Hence Warner (1965) concluded that

$$\hat{\pi}_{AW} = \frac{P-1}{2P-1} + \frac{n_1}{(2P-1)n} = \frac{\hat{\lambda}-(1-P)}{(2P-1)}$$

is the maximum likelihood estimator (MLE) for π_A. But this conclusion is erroneous (1) because an MLE for a parameter must be an element in the parametric space, but (2) $\hat{\pi}_{AW}$ does not necessarily belong to the parametric space $[0, 1]$ of π_A.

The true MLE for π_A is $\tilde{\pi}_{AW}$ given by

$$\begin{aligned} \tilde{\pi}_{AW} &= 1 && \text{if } \hat{\lambda} \geq P > \tfrac{1}{2} \\ &= 0 && \text{if } \hat{\lambda} \leq (1-P) < \tfrac{1}{2} \\ &= \hat{\pi}_{AW} && \text{if } 1-P < \hat{\lambda} < P, P > \tfrac{1}{2} \\ &= \hat{\pi}_{AW} && \text{if } P < \hat{\lambda} < 1-P, P < \tfrac{1}{2} \\ &= 1 && \text{if } \hat{\lambda} \leq P < \tfrac{1}{2} \\ &= 0 && \text{if } \hat{\lambda} \geq (1-P) > \tfrac{1}{2} \end{aligned}$$

Here $\hat{\lambda} = n_1/n$. Since $\lambda \in [1-P, P]$ if $P > 1/2$ or $\lambda \in [P, 1-P]$ if $P < 1/2$, this $\hat{\lambda}$ is not the MLE of λ. The true MLE, say $\tilde{\lambda}$, has to be obtained by a truncation process similar to that described above. Devore (1977) and, independently, Flinger *et al.* (1977) and Singh (1976), have pointed out the above-noted snag in Warner's (1965) claim for $\hat{\pi}_{AW}$ as the MLE of π_A and obtained the true MLE as above.

This MLE is not an unbiased estimator for π_A. So its mean square error (MSE) about π_A should be used as a measure of its error. Since $\tilde{\pi}_{AW}$ is obtained by truncating $\hat{\pi}_{AW}$ at the two endpoints of its range, obviously,

$$|\tilde{\pi}_{AW} - \pi_A| < |\hat{\pi}_{AW} - \pi_A|$$

implying that

$$\text{MSE}(\tilde{\pi}_{AW}) < \text{MSE}(\hat{\pi}_{AW}) = \text{var}(\hat{\pi}_{AW})$$

It is instructive for a reader to find a formula for $\text{MSE}(\tilde{\pi}_{AW})$ and then verify this directly.

Devore (1977) suggests a procedure for obtaining a UE which is also the MLE for π_A by slightly changing the RRT from the one due to Warner as follows. Suppose that n is even and a sample of n persons is taken but that each sample person chosen is asked to choose without replacement one card each from a pack of n cards, onehalf of which demand an unconditional "yes" reply and the other half a truthful "yes" or "no" reply about the person's bearing the stigmatizing character A. Obviously, if X denotes the observed number of "yes" replies, $(X - n/2)$ replies are made independently of each other, and in fact $Y = (X - n/2)$ has the binomial distribution with parameters $n/2$ and π_A. So

$$\hat{\pi}_A = \frac{2Y}{n} = \frac{2(X - n/2)}{n} = \frac{2X}{n} - 1$$

gives a UE for π_A and also it is the MLE for π_A. Since $n/2 \le X \le n$, there can be no anomaly and $\hat{\pi}_A$ takes values only within $[0, 1]$. The variance of $\hat{\pi}_A$ is

$$\text{var}(\hat{\pi}_A) = \frac{2}{n}\pi_A(1 - \pi_A)$$

with a UE as

$$\hat{\text{var}}(\hat{\pi}_A) = \frac{2}{n-2}\hat{\pi}_A(1 - \hat{\pi}_A)$$

A1.4 SIMPLE RANDOM SAMPLING WITHOUT REPLACEMENT (SRSWOR) AND THE WARNER MODEL

Kim and Flueck (1978a) consider both SRSWR and SRSWOR in n draws from a finite population of N persons and envisage an application of RRT in either an independent or a dependent manner as follows.

Suppose that a box contains M identical-looking cards with a proportion P of them of type A asking about bearing the stigmatizing character A and the complementary proportion $(1 - P)$ of them of type $\bar{A}$ asking about bearing $\bar{A}$. These authors consider the possibilities of each person selected choosing one card with or without replacement. So they consider four schemes, I to IV:

I. Sample selection WR, card selection WR
II. Sample selection WOR, card selection WR
III. Sample selection WR, card selection WOR
IV. Sample selection WOR, card selection WOR

Following Boruch (1972), Kim and Flueck (1978a) show that the Warner estimator $\hat{\pi}_{AW}$ is unbiased for π_A if $P \neq 1/2$ for each of the schemes. The respective variances of $\hat{\pi}_{AW}$ under the four schemes are as follows:

$$V_1 = \frac{\pi_A(1 - \pi_A)}{n} + \frac{P(1 - P)}{n(2P - 1)^2}$$

$$V_2 = \frac{\pi_A(1 - \pi_A)}{n}\frac{N - n}{N - 1} + \frac{P(1 - P)}{n(2P - 1)^2}$$

$$V_3 = \frac{\pi_A(1-\pi_A)}{n} + \frac{4\pi_A(1-\pi_A)P(1-P)}{n(2P-1)^2}\left(1-\frac{M-n}{M-1}\right) + \frac{P(1-P)}{n(2P-1)^2}\frac{M-n}{M-1}$$

$$V_4 = \frac{\pi_A(1-\pi_A)}{n(2P-1)^2}\left\{(4P^2-4P)\left[\frac{N(M-n)-n(M-1)}{(N-1)(M-1)}\right]+\frac{N-n}{N-1}\right\} + \frac{P(1-P)}{n(2P-1)^2}\frac{M-n}{M-1} \qquad \left(P \neq \frac{1}{2}\right)$$

Further, it was shown by Kim and Flueck (1978a) that

1. $V_1 > V_i (i=2, 3, 4)$
2. $V_2 \geq V_4$ if $4\pi_A(1-\pi_A) \leq (N-1)/N$
3. $V_3 \geq V_4$ if $4P(1-P) \leq (M-1)/M$

In passing we may note that a UE of V_2 is given by

$$\hat{V}_2 = \frac{(N-n)\hat{\pi}_{AW}(1-\hat{\pi}_{AW})+[(N-1)P(1-P)/(2P-1)^2]}{N(n-1)}$$

Kim and Flueck (1978a) also suggest alternative procedures for obtaining randomized responses in interdependent ways. For example, one may start with a box of M beads of two colors in proportions $P/(1-P)$ and ask successive sample persons to choose one bead each at random and retain it as a memento in recognition of his or her cooperation. A second procedure might be to use a big plastic board with M identical-looking paper strips attached and to ask each sample person to tear off one of the strips and follow the instruction written on the reverse (concealed) side to state "yes" or "no" as to bearing A or $\bar{A}$. However, the authors overlook the fact that people may hesitate to cooperate because they may not believe that there are really strips with two types of complementary instructions in acceptably correct proportions.

A1.5 AUGMENTATION MODELING

Let us associate a variable C to a person so as to assume the value 1(0) when he or she bears A($\bar{\text{A}}$). Let an augmentation variable X, distributed independently of C, be chosen taking values 0 and 1 with probabilities P and $1-P$, respectively. Let $Y=C+X$ and a selected person be asked to respond R such that

$$R = Y \qquad \text{if } Y \leq 1$$
$$= Y - 2 \qquad \text{if } Y > 1$$

Then R assumes the value 1 with probability

$$\lambda = \pi_A(2P - 1) + (1 - P)$$

and 0 with probability $1 - \lambda$.

Hence this RRT is equivalent to Warner's original RRT, yielding the same estimator $\hat{\pi}_{AW}$ for π_A with the same variance. But this augmentation modeling has a potentiality to be exploited later when we cover polychotomous populations in Chapter 3, following the work of Kim and Flueck (1978b).

It may be of interest to note that in Warner's model it is tacitly assumed that the process of selection of a question, relating to membership in A or $\bar{A}$, through an RR experiment, is independent of a respondent's true characteristic (cf. the fact that X is independent of C in the context of the augmentation model). But this may not always be tenable in practice. If, for example, $P > 1/2$ and A carries a stigma whereas $\bar{A}$ does not, a respondent may feel safe to announce "no" rather than "yes", irrespective of his or her true characteristic and the outcome of the randomization experiment.

It is possible in such a situation that Warner's RRT or, equivalently, the augmentation device may fail because of probable falsity in responses. As a way out, Takahasi and Sakasegawa (1977) suggest what may be called an indirect response technique which avoids strict randomization in eliciting a response. We describe their scheme in Section 2.7. Here we briefly note their suggestion. Taking, for example, three independent SRSWR's S_1, S_2, and S_3, say, a person in each sample is to choose arbitrarily one of three alternatives, I, II, and III (say, three flowers). If the choice is I, a person bearing A($\bar{A}$) is to report 1(0), 1(0), or 0(1), respectively, if he or she is in S_1, S_2, or S_3. For the choices II and III, respectively, the reports are, correspondingly, 0(1), 1(0), 1(0) and 1(0), 0(1), 1(0). Defining λ_i as the probability of response 1 from a person in the ith sample, these authors work out

$$\pi_A = \sum_1^3 \lambda_i - 1$$

and hence, using the sample proportions of replies, they obtain a UE, say $\hat{\pi}_{ATS}$, for π_A.

Incidentally, Singh (1978), following Flinger *et al.* (1977), points out that this UE is not the MLE and the true MLE may be obtained by an appropriate truncation procedure. They also show that the latter naturally has a smaller MSE about π_A than var ($\hat{\pi}_{ATS}$).

If, of course, both A and $\bar{A}$ are stigmatizing, for example, when A denotes voting for a particular political party, a respondent may not have any feeling of safety in replying "no" rather than "yes" and the problem referred to here may not arise. In this context it may be pertinent to refer to the concept of "symmetry of response" introduced by Bourke (1974a, b). A response is symmetric if it, of itself, does not convey any message about the true characteristic of a respondent. In case an RRT appears to lack such symmetry, Bourke (1974a, b) suggests a randomization in response a second time (i.e., a two-stage randomization) as a procedure to promote a higher sense of security about the protection of the respondent's privacy despite a truthful statement regarding the trait. In succeeding chapters we illustrate the application of these concepts and proposals from Bourke. Here they are not greatly relevant because Warner's original "yes-no" response scheme, which may refer to both A and $\bar{A}$, has the symmetry property and may not fail in practice even if $\bar{A}$ is innocuous and only A is sensitive.

EXERCISES

A1. Work out the MSE of the MLE for π_A under the Warner model.

A2. Find the MLE for π_A in scheme II of Kim and Flueck (1978a), and obtain its MSE.

A3. Obtain the MLE for π_A for the scheme of Takahasi and Sakasegawa (1977) illustrated above.

REFERENCES

Boruch, R. F. (1972). Relations among statistical methods for assuring confidentiality of social research data. *Soc. Sci. Res.* **1**, 403–414.

Bourke, P. D. (1974a). Symmetry of response in RR designs. *Tech. Rep.* 75, Institute of Statistics, University of Stockholm, Sweden.

Bourke, P. D. (1974b). Vector response in RR designs. Private communication.

Devore, J. L. (1977). A note on the RR techniques. *Comm. Statist.—Theory Methods* **6**, 1525–1529.

Flinger, M. A., Policello, G. E. and Singh, J. (1977). A comparison of two RR survey methods with consideration for the level of respondent protection. *Comm. Statist.—Theory Methods* **6**, 1511–1524.

Kim, Jong-Ik, and Flueck, John A. (1978a). Modifications of the randomized response technique for sampling without replacement. *Proc. ASA. Sec. Surv. Res. Methods*, 346–350.

Kim, Jong-Ik, and Flueck, John A. (1978b). An additive randomized response model. *Proc. ASA. Sec. Surv. Res. Methods*, 351–355.

Moors, J. J. A. (1985). Estimation in truncated parameter spaces. Ph.D. thesis, Kathalieke Hogeschool, Tilburg.

Raghavarao, D. (1978). On an estimation problem in Warner's RR technique. *Biometrics* **34**, 87–90.

Singh, J. (1976). A note on RR techniques. *Proc. ASA. Soc. Statist. Sec.*, 772.

Singh, J. (1978). A note on maximum likelihood estimation from randomized response models. *Proc. ASA. Soc. Statist. Sec.*, 282–283.

Takahasi, K. and Sakasegawa, H. (1977). An RR technique without use of any randomizing device. *Ann. Inst. Statist Math.* **29**, 1–8.

Warner, S. L. (1965). RR: a survey technique for eliminating evasive answer bias. *J. Amer. Statist. Assoc.* **60**, 63–69.

2

The Unrelated-Question Model

2.1. INTRODUCTION

Consider, as in Chapter 1, the problem of estimating the proportion π_A of population members of a sensitive group A. Closer examination reveals the following distinctive feature of the Warner model as applied to such a situation. Essentially, each respondent is supplied with two questions, of which he or she selects one at random and replies only to the question chosen. In the process the respondent's privacy is protected since the interviewer is unaware of the question to which his or her response relates. In order that the procedure be relevant for the basic problem, that is, estimation of π_A, one of these two questions should relate to membership in A. In the Warner model, the other question relates to membership in the complementary group $\bar{A}$. Thus the two questions relate to groups that are perfectly negatively associated.

A little reflection, however, makes it intuitively evident that to protect the confidentiality of the respondent it is not quite necessary that the two questions relate to such complementary or negatively associated groups. In fact, it is enough that the two groups are unrelated. This possibility leads to the unrelated-question model, which was suggested by Simmons as recorded by Greenberg *et al.* (1969).

In this approach, instead of requesting that the respondent state on a probability basis whether he or she belongs to group A or the complementary group $\bar{A}$, the respondent is asked to answer "yes" or "no" to queries regarding belonging to group A or to another group, Y. Membership in group Y is unrelated, in the sense of statistical association, to that in group A. As before, queries about A and Y are chosen with complementary probabilities. Apart from being unrelated to group A, group Y should be such that both Y and its complement $\bar{Y}$ are innocuous. Only in that case will the respondent feel that a "yes" or "no" reply will not reveal his or her identity. In estimating the proportion of drug addicts in a population, each interviewee may, for example, be requested to select and answer one of two questions "Do you habitually consume illicit drugs?" or "Do you like the game of cricket?" Since the second question relates to an innocuous characteristic that is unrelated to drug addiction, one can reasonably expect that the respondent's privacy is being sufficiently well protected.

2.2. THE CASE OF KNOWN π_Y

Consider first the simple case when π_Y, the true proportion in group Y in the population, is known. An SRSWR of size n is drawn from the population and each interviewee is asked to report only "yes" or "no" regarding belonging to A (chosen with probability P) or to Y (chosen with probability $1-P$). As in Chapter 1, a spinner or deck of cards or any other suitable randomization device may be employed for this purpose. Evidently, the probability of a "yes" response is

$$\lambda = P\pi_A + (1-P)\pi_Y \tag{2.1}$$

If n_1 be the number of "yes" responses in the sample and $\hat{\lambda} = n_1/n$, a UE of π_A is

$$\hat{\pi}_{AU1} = \frac{\hat{\lambda} - (1-P)\pi_Y}{P} \tag{2.2}$$

with

$$\operatorname{var}(\hat{\pi}_{AU1}) = \frac{\lambda(1-\lambda)}{(nP^2)} \tag{2.3}$$

which can be estimated unbiasedly by

$$\hat{\operatorname{var}}(\hat{\pi}_{AU1}) = \frac{\hat{\lambda}(1-\hat{\lambda})}{(n-1)P^2}$$

It may be remarked that the assumption that π_Y is known is not altogether unrealistic. In this connection, Horvitz *et al.* (1976) mention an interesting observation due to R. Morton that knowledge of π_Y can always be achieved by incorporating it in the randomization device. For example, in the device there may be three statements: (1) the sensitive statement, (2) an instruction to say "yes", and (3) an instruction to say "no", to be chosen with respective probabilities P_1, P_2, P_3, where $P_1+P_2+P_3=1$. Then the probability of a "yes" response is

$$\lambda = P_1\pi_A + P_2$$

which can, as well, be represented in the form (2.1) with $P=P_1$, $\pi_Y = P_2/(P_2+P_3)$. In other words, the procedure automatically induces an unrelated character Y, with $\pi_Y = P_2/(P_2+P_3)$, which may be chosen arbitrarily. An equivalent procedure has also been suggested by Boruch (1972), who calls it a "contamination design."

Although the example above is conceptually elegant in terms of its applicability to real-life problems, difficulties may arise because of its essentially theoretical nature. In particular, in many practical situations it may be difficult to convince the respondent regarding the protection of his or her privacy using such an artificially constructed unrelated character with known π_Y. Then one has to develop the procedure with a more concrete and realistic unrelated character for which π_Y is quite likely to be unknown. The details of such a situation are presented in the following section.

2.3. THE CASE OF UNKNOWN π_Y

So far there has been only one unknown parameter, π_A, and for its estimation applying an appropriate RR technique, a single sample was seen to be sufficient. Now that there are two unknown parameters, π_A and π_Y, it is intuitively felt that at least two samples will be required. This is easy to appreciate, particularly by observing that a single relation such as (2.1) can no longer yield an estimator of π_A since π_Y is there as a nuisance parameter in addition to π_A. So in the case of unknown π_Y, two independent SRSWRs of sizes n_1 and n_2 may be drawn from the population. Then the randomization device is such that each respondent in the ith sample answers "yes" or "no" about A with probability P_i (and about Y with $1-P_i$), $i=1, 2$. Clearly, the probability of a "yes" response in the ith sample is

$$\lambda_i = P_i\pi_A + (1-P_i)\pi_Y, \qquad (i=1,2) \tag{2.4}$$

Let n_{i1} be the number of "yes" responses in the ith sample and $\hat{\lambda}_i = n_{i1}/n_{i\cdot}$. Then taking $P_1 \neq P_2$ from the observation that

$$E(\hat{\lambda}_i) = \lambda_i = P_i\pi_A + (1-P_i)\pi_Y, \qquad (i=1,2)$$

a UE of π_A works out as

$$\hat{\pi}_{AU2} = \frac{\hat{\lambda}_1(1-P_2) - \hat{\lambda}_2(1-P_1)}{P_1 - P_2} \tag{2.5}$$

Since $\mathrm{var}(\hat{\lambda}_i) = \lambda_i(1-\lambda_i)/n_i$ and $\hat{\lambda}_1$ and $\hat{\lambda}_2$ are independent, one obtains

$$\mathrm{var}(\hat{\pi}_{AU2}) = \left[\frac{(1-P_2)^2\lambda_1(1-\lambda_1)}{n_1} + \frac{(1-P_1)^2\lambda_2(1-\lambda_2)}{n_2}\right] \Big/ (P_1 - P_2)^2 \tag{2.6}$$

Analogously to (1.4), $\mathrm{var}(\hat{\pi}_{AU2})$ can be estimated unbiasedly by

$$\hat{\mathrm{var}}(\hat{\pi}_{AU2}) = \left[\frac{(1-P_2)^2\hat{\lambda}_1(1-\hat{\lambda}_1)}{n_1 - 1} + \frac{(1-P_1)^2\hat{\lambda}_2(1-\hat{\lambda}_2)}{n_2 - 1}\right] \Big/ (P_1 - P_2)^2$$

In an earlier work, Horvitz *et al.* (1967) considered an equivalent single-sampling scheme with two trials per respondent and two different selection probabilities for the sensitive character when π_Y is unknown.

2.4. OPTIMAL CHOICE OF DESIGN PARAMETERS

For successful application of the unrelated-question scheme, keeping efficiency and protection of privacy in mind, one has to select the unrelated characteristic Y and the design parameters n_1, n_2, P_1, P_2 in appropriate manners. In this connection, the recommendations of Greenberg *et al.* (1969) and Moors (1971) will be described first.

In practical situations, the total sample size $n\,(=n_1+n_2)$ is fixed from a consideration of available resources. One is then concerned with the selection of n_1, n_2 so that $\mathrm{var}(\hat{\pi}_{AU2})$, as in (2.6), is minimized subject to $n_1 + n_2 = n$. A solution to this constrained optimization problem is available by observing, through a simple application of the Cauchy-Schwarz inequality, that

$$\left[\frac{(1-P_2)^2\lambda_1(1-\lambda_1)}{n_1} + \frac{(1-P_1)^2\lambda_2(1-\lambda_2)}{n_2}\right](n_1 + n_2)$$
$$\geq \{(1-P_2)[\lambda_1(1-\lambda_1)]^{1/2} + (1-P_1)[\lambda_2(1-\lambda_2)]^{1/2}\}^2 \tag{2.7}$$

with equality if and only if

$$\frac{n_1}{n_2} = \left[\frac{(1-P_2)^2\lambda_1(1-\lambda_1)}{(1-P_1)^2\lambda_2(1-\lambda_2)}\right]^{1/2} \tag{2.8}$$

By (2.6) and (2.7), the optimal choice of n_1 and n_2 for minimizing $\text{var}(\hat{\pi}_{AU2})$ subject to $n_1+n_2=n$ must satisfy (2.8), and for this allocation (2.6) reduces to

$$\text{var}(\hat{\pi}_{AU2})=\left\{\frac{(1-P_2)[\lambda_1(1-\lambda_1)]^{1/2}+(1-P_1)[\lambda_2(1-\lambda_2)]^{1/2}}{n^{1/2}(P_1-P_2)}\right\}^2 \quad (2.9)$$

It may be noted that since λ_1 and λ_2 are actually unknown, information regarding them as obtained from past experience or a pilot survey may be helpful to realize (2.8) in practical applications.

Next, to choose the unrelated characteristic Y, or equivalently π_Y, so as to minimize (2.9), observe that π_Y occurs only in λ_1 and λ_2 in (2.9). Since $[\lambda_i(1-\lambda_i)]^{1/2}$ is maximum at $\lambda_i=1/2$, is symmetric about 1/2, and is concave, one has to take λ_i as remote from 1/2 as possible. By (2.4), this may be achieved by choosing Y such that π_Y is on the same side of 1/2 as is π_A and $|\pi_Y-1/2|$ is maximized. If one has no idea whatsoever on which side of 1/2 to expect π_A, a moderate value of π_Y between 0.25 and 0.75 will be helpful.

Turning to the problem of selection of P_1 and P_2, Greenberg *et al.* (1969) suggest a choice of P_1 near 0.20 ± 0.10 or 0.80 ± 0.10 and a choice of P_2 as $P_2=1-P_1$. Although this recommendation is based essentially on heuristic considerations, Moors (1971) suggests a more systematic approach to the problem as follows. Since $P_1\neq P_2$ so that π_A is estimable, assume, without loss of generality, that $P_1>P_2$. Differentiating (2.9) with respect to P_1, one finds after some simplification, using (2.4), that the sign of $\partial\,\text{var}(\hat{\pi}_{AU2})/\partial P_1$ is the same as that of

$$(\tfrac{1}{2}-\lambda_1)(P_1-P_2)(\pi_A-\pi_Y)[\lambda_1(1-\lambda_1)]^{-1/2}-[\lambda_1(1-\lambda_1)]^{1/2} \\ -[\lambda_2(1-\lambda_2)]^{1/2}$$

Since by (2.4), $(P_1-P_2)(\pi_A-\pi_Y)=\lambda_1-\lambda_2$, the expression above equals

$$-\tfrac{1}{2}[\lambda_1(1-\lambda_1)]^{-1/2}\{[\lambda_1(1-\lambda_2)]^{1/2}+[\lambda_2(1-\lambda_1)]^{1/2}\}^2$$

which certainly cannot be positive. Also, $\partial\,\text{var}(\hat{\pi}_{AU2})/\partial P_2\geq 0$. It follows that to minimize (2.9) one should take $P_2=0$ and P_1 as large as practicable. The choice $P_2=0$ is permissible, demanding that a respondent in the second sample divulge the truth about Y which is innocuous. Apart from minimizing (2.9), the choice $P_2=0$ also has administrative and other advantages. For example, with $P_2=0$, the randomization device needs to be explained only to the individuals in the first sample, not to those in the second sample, and consequently, the interview costs are reduced.

Lanke (1975) also considers the problem of choosing π_Y assuming that A is stigmatizing but not $\bar{A}$ and further that $\pi_A < 1/2$. Thus he claims that only a "yes" response is embarrassing and that the more so that it is, the less often the unrelated question is answered "yes". So even a choice of Y with $\pi_Y = 1$ is acceptable. For Moors' optimal version with $P_2 = 0$, the expression (2.9) reduces to

$$\operatorname{var}(\hat{\pi}_{AU2}) = \frac{\{[\lambda_1(1-\lambda_1)]^{1/2} + (1-P_1)[\pi_Y(1-\pi_Y)]^{1/2}\}^2}{nP_1^2} \tag{2.10}$$

which Lanke (1975) treats as a function of π_Y for $\pi_Y \in (\pi_-, \pi_+)$, where $\pi_- < 1/2$, $\pi_+ > 1/2$, and $\pi_- + \pi_+ > 1$. He observes that (1) this function is concave, (2) it attains by (1) a minimum at either π_- or π_+, and (3) if π_+ is sufficiently greater than π_-, the minimum will be attained at π_+ or else at π_-. Lanke also treats the question of protection of privacy, which we take up in Chapter 5.

2.5. COMPARISON OF THE WARNER MODEL AND THE UNRELATED-QUESTION MODEL

It is interesting to compare the efficiencies of the estimators of π_A under the Warner and unrelated-question models. Dowling and Shachtman (1975) made some analytic investigations in this connection and obtained the following results. The first of these deals with the unrelated-question model with known π_Y, the second one covers the situation when π_Y is unknown.

Theorem 2.1. Suppose that π_Y is known and let $P \in (p_0, 1)$, where $p_0 = 0.339333\ldots$ is the unique solution of

$$(1+P^2)^{-1} = 4P(1-P) \tag{2.11}$$

in $[0, 1/2]$. Then var $(\hat{\pi}_{AU1})$ as in (2.3) is less than $\operatorname{var}(\hat{\pi}_{AW})$ as in (1.3) for all π_A, π_Y, $0 \leq \pi_A, \pi_Y \leq 1$.

Proof. Since $(1+P^2)^{-1}$ is strictly decreasing from 1 to 4/5 and $4P(1-P)$ is strictly increasing from 0 to 1 in $[0, 1/2]$, equation (2.11) has a unique solution, say $p_0 \in [0, 1/2]$. The approximate value of p_0 may be seen, by numerical methods, to be as stated above.

Now, by (2.1) and (2.3), working through a few steps which we do not show, we obtain

$$\operatorname{var}(\hat{\pi}_{AU1})=\frac{\pi_A(1-\pi_A)}{n}+\frac{1-P}{nP^2}K_P(\pi_Y,\pi_A) \tag{2.12}$$

where

$$K_P(\pi_Y,\pi_A)=P(1-2\pi_Y)\pi_A+\pi_Y[1-(1-P)\pi_Y]$$

By (1.3) and (2.12), $\operatorname{var}(\hat{\pi}_{AU1})<\operatorname{var}(\hat{\pi}_{AW})$ for all π_Y, π_A $(0\le\pi_Y,\pi_A\le 1)$ if and only if

$$\max_{0\le\pi_Y,\,\pi_A\le 1} K_P(\pi_Y,\pi_A)<\frac{P^3}{(2P-1)^2} \tag{2.13}$$

To derive the left-hand member of (2.13), note that $K_P(\pi_Y,\pi_A)=K_P(1-\pi_Y, 1-\pi_A)$, so that it is enough to maximize over $0\le\pi_Y\le 1/2, 0\le\pi_A\le 1$. Also, for a fixed $\pi_Y\in[0, 1/2]$, $K_P(\pi_Y,\pi_A)$ is nondecreasing in π_A. These observations, together with some elementary calculus, yield

$$\begin{aligned}
&\max_{0\le\pi_Y,\,\pi_A\le 1} K_P(\pi_Y,\pi_A)\\
&\quad=\max_{0\le\pi_Y\le 1/2}\ \max_{0\le\pi_A\le 1} K_P(\pi_Y,\pi_A)=\max_{0\le\pi_Y\le 1/2} K_P(\pi_Y,1)\\
&\quad=\max_{0\le\pi_Y\le 1/2}[P+(1-2P)\pi_Y-(1-P)\pi_Y^2]\\
&\quad=\begin{cases}[4(1-P)]^{-1} & \text{if } P\in(0,\tfrac{1}{2})\\ P & \text{if } P\in(\tfrac{1}{2},1)\end{cases}
\end{aligned} \tag{2.14}$$

In (2.14), the case $P=1/2$ has been excluded since then $\hat{\pi}_{AW}$ becomes undefined. From (2.13) and (2.14), $\operatorname{var}(\hat{\pi}_{AU1})<\operatorname{var}(\hat{\pi}_{AW})$ for all π_Y, π_A $(0\le\pi_Y,\pi_A\le 1)$ if either (a) $P\in(0, 1/2)$ and $[4(1-P)]^{-1}<P^3/(2P-1)^2$, or (b) $P\in(1/2, 1)$ and $P<P^3/(2P-1)^2$. The inequality in (b) holds for all $P\in(1/2, 1)$, while that in (a) is equivalent to $(1+P^2)^{-1}<4P(1-P)$ and hence holds for $P\in(p_0, 1/2)$. This completes the proof.

In the case of unknown π_Y, Dowling and Shachtman (1975) consider Moors' (1971) optimized version of the unrelated question model with $P_2=0$, $P_1=P$. Then by (2.10),

$$\operatorname{var}(\hat{\pi}_{AU2})=\frac{\{[\lambda_1(1-\lambda_1)]^{1/2}+(1-P)\,[\pi_Y(1-\pi_Y)]^{1/2}\}^2}{nP^2} \tag{2.15}$$

where

$$\lambda_1 = P\pi_A + (1-P)\pi_Y \tag{2.16}$$

Theorem 2.2. Let $P \in (p_{00}, 1)$, where $p_{00} = (1/2)(3-\sqrt{5})$. Then $\operatorname{var}(\hat{\pi}_{AU2})$ as in (2.15) is less than $\operatorname{var}(\hat{\pi}_{AW})$ as in (1.3) for all $\pi_A, \pi_Y, 0 \leq \pi_A, \pi_Y \leq 1$.

Proof. By elementary considerations and making use of (2.16),

$$\begin{aligned}
&\{[\lambda_1(1-\lambda_1)]^{1/2} + (1-P)[\pi_Y(1-\pi_Y)]^{1/2}\}^2 \\
&\leq [\lambda_1 + (1-P)(1-\pi_Y)][1-\lambda_1 + (1-P)\pi_Y] \\
&= P^2\pi_A(1-\pi_A) + (1-P)
\end{aligned}$$

with equality if and only if $\pi_Y = (1-P\pi_A)/(2-P)$. Hence by (2.15), for fixed $\pi_A (0 \leq \pi_A \leq 1)$,

$$\max_{0 \leq \pi_Y \leq 1} \operatorname{var}(\hat{\pi}_{AU2}) = \frac{\pi_A(1-\pi_A)}{n} + \frac{1-P}{nP^2} \tag{2.17}$$

Comparing the above with (1.3), it follows that $\operatorname{var}(\hat{\pi}_{AU2}) < \operatorname{var}(\hat{\pi}_{AW})$ holds for all $\pi_A, \pi_Y (0 \leq \pi_A, \pi_Y \leq 1)$ provided that

$$\frac{1-P}{P^2} < \frac{P(1-P)}{(2P-1)^2}$$

or

$$P^3 - 4P^2 + 4P - 1 > 0 \tag{2.18}$$

The roots of the cubic in (2.18) are 1, $(1/2)(3-\sqrt{5}) = p_{00}$, and $(1/2)(3+\sqrt{5}) > 1$. Hence given that $0 < P < 1$, (2.18) holds if $P \in (p_{00}, 1)$. This completes the proof.

The proof of Theorem 2.2 is essentially based on elementary considerations and is different from that in Dowling and Shachtman (1975), who employ differential calculus in obtaining (2.17). Prior to Dowling and Shachtman's work, Moors (1971) also compared (2.15) with (1.3) in the special case $\pi_Y = 1/2$ and observed that (2.15) is less than (1.3) provided that $P > 1/2$ (see Exercise 2.1). This finding of Moors, of course, follows as a special case of Theorem 2.2.

The foregoing comparison between the Warner model and the unrelated-question model is framed in terms of the efficiencies of the estimators of π_A. A more meaningful comparison should also take into account the question of protection of privacy. The relevant results in this regard are presented in Chapter 5.

2.6. MODEL WITH TWO UNRELATED CHARACTERS

Folsom *et al.* (1973) consider an unrelated-question model with two nonsensitive attribute characteristics, Y_1 and Y_2, in conjunction with a dichotomous sensitive character A. Suppose that π_{Y_1} and π_{Y_2}, the true proportions of Y_1 and Y_2, are unknown. Two independent SRSWRs of sizes n_1 and n_2 are drawn. The respondents in both the samples answer a direct question on a nonsensitive topic and also one of two questions selected by a randomizing device. The scheme is shown in Table 2.1. In both samples let the sensitive question A be selected with probability P, and for $i=1, 2$, let $\lambda_i^r(\lambda_i^d)$ be the probability of a "yes" response to the question selected by RR(DR) in the ith sample. Then

$$\lambda_1^r = P\pi_A + (1-P)\pi_{Y_1} \tag{2.19a}$$

$$\lambda_2^r = P\pi_A + (1-P)\pi_{Y_2} \tag{2.19b}$$

$$\lambda_1^d = \pi_{Y_2} \tag{2.19c}$$

$$\lambda_2^d = \pi_{Y_1} \tag{2.19d}$$

Let $\hat{\lambda}_1^r$, $\hat{\lambda}_2^r$, $\hat{\lambda}_1^d$, and $\hat{\lambda}_2^d$ denote the usual UEs of λ_1^r, λ_2^r, λ_1^d, and λ_2^d, respectively, given by the corresponding sample proportions. Then from (2.19a, d) a UE of π_A is obtained as

$$\hat{\pi}_A(1) = \frac{\hat{\lambda}_1^r - (1-P)\hat{\lambda}_2^d}{P} \tag{2.20a}$$

TABLE 2.1

Technique used with respondents	Sample 1	Sample 2
Randomized response (RR)	Question A Question Y_1	Question A Question Y_2
Direct response (DR)	Question Y_2	Question Y_1

and similarly, (2.19b, c) lead to another UE of π_A as

$$\hat{\pi}_A(2) = \frac{\hat{\lambda}_2^r - (1-P)\hat{\lambda}_1^d}{P} \tag{2.20b}$$

In practice, one would like to use an optimal linear combination of these two estimators, $\hat{\pi}_A(1)$ and $\hat{\pi}_A(2)$, which will be unbiased for π_A and will have the minimum variance among all such unbiased linear combinations. This consideration leads to the following result.

Theorem 2.3. Let σ_{11} and σ_{22} denote the variances of $\hat{\pi}_A(1)$ and $\hat{\pi}_A(2)$ and let σ_{12} denote their covariance. Then the best linear unbiased estimator of π_A is

$$\hat{\pi}_{AF} = w\,\hat{\pi}_A(1) + (1-w)\,\hat{\pi}_A(2)$$

with

$$\text{var}(\hat{\pi}_{AF}) = \frac{\sigma_{11}\sigma_{22} - \sigma_{12}^2}{\sigma_{11} + \sigma_{22} - 2\sigma_{12}} \tag{2.21}$$

where $w = (\sigma_{22} - \sigma_{12})/(\sigma_{11} + \sigma_{22} - 2\sigma_{12})$.

Proof. A typical linear combination of $\hat{\pi}_A(1)$ and $\hat{\pi}_A(2)$ is of the form $w_1\hat{\pi}_A(1) + w_2\hat{\pi}_A(2)$, which is unbiased for π_A if and only if $w_1 + w_2 = 1$. With $w_2 = 1 - w_1$,

$$\begin{aligned} &\text{var}[w_1\hat{\pi}_A(1) + (1-w_1)\hat{\pi}_A(2)] \\ &= w_1^2\sigma_{11} + (1-w_1)^2\sigma_{22} + 2w_1(1-w_1)\sigma_{12} \end{aligned} \tag{2.22}$$

and a simple application of differential calculus shows that the above, as a function of w_1, is minimum when $w_1 = w$. If $w_1 = w$, a routine computation reduces (2.22) to (2.21), completing the proof.

To obtain an explicit expression for $\hat{\pi}_{AF}$, it is important to find σ_{11}, σ_{22}, and σ_{12}. Since the two samples are independent, one has

$$\sigma_{11} = P^{-2}\left[\frac{\lambda_1^r(1-\lambda_1^r)}{n_1} + \frac{(1-P)^2\,\pi_{Y_1}(1-\pi_{Y_1})}{n_2}\right] \tag{2.23a}$$

$$\sigma_{22} = P^{-2}\left[\frac{\lambda_2^r(1-\lambda_2^r)}{n_2} + \frac{(1-P)^2\,\pi_{Y_2}(1-\pi_{Y_2})}{n_1}\right] \tag{2.23b}$$

Also, denoting covariance by "cov," it follows by (2.20a, b) that

$$\begin{aligned}\sigma_{12} &= -(1-P)P^{-2}[\text{cov}(\hat{\lambda}_1^r, \hat{\lambda}_1^d) + \text{cov}(\hat{\lambda}_2^r, \hat{\lambda}_2^d)] \\ &= -(1-P)P^{-2}\left(\frac{\lambda_1^{rd} - \lambda_1^r \pi_{Y_2}}{n_1} + \frac{\lambda_2^{rd} - \lambda_2^r \pi_{Y_1}}{n_2}\right)\end{aligned} \tag{2.23c}$$

where λ_i^{rd} is the probability of a "yes" response to both questions in sample i ($i = 1, 2$). Defining

π_{AY_i} = true proportion of individuals belonging to both A and Y_i ($i=1, 2$)

$\pi_{Y_1Y_2}$ = true proportion of individuals belonging to both Y_1 and Y_2

it is easy to see that

$$\lambda_1^{rd} = P\pi_{AY_2} + (1-P)\pi_{Y_1Y_2} \qquad \lambda_2^{rd} = P\pi_{AY_1} + (1-P)\pi_{Y_1Y_2} \tag{2.24}$$

In Exercise 2.2 we ask the reader to verify (2.23c) and (2.24).

Of course, one cannot use the estimator $\hat{\pi}_{AF}$ in the form in which it has been presented in Theorem 2.3. This is because $\hat{\pi}_{AF}$ depends on w, which involves the unknown parameters σ_{11}, σ_{22}, and σ_{12}. In practice, one should, therefore, first estimate σ_{11}, σ_{22}, and σ_{12} from (2.23a, b, c), replacing λ_i^r, λ_i^{rd}, and π_{Y_i} ($i=1, 2$) by the corresponding sample proportions. By (2.19c, d) and (2.20a, b, c), these estimators are

$$\begin{aligned}\hat{\sigma}_{11} &= P^{-2}\left[\frac{\hat{\lambda}_1^r(1-\hat{\lambda}_1^r)}{n_1} + \frac{(1-P)^2\,\hat{\lambda}_2^d(1-\hat{\lambda}_2^d)}{n_2}\right] \\ \hat{\sigma}_{22} &= P^{-2}\left[\frac{\hat{\lambda}_2^r(1-\hat{\lambda}_2^r)}{n_2} + \frac{(1-P)^2\,\hat{\lambda}_1^d(1-\hat{\lambda}_1^d)}{n_1}\right] \\ \hat{\sigma}_{12} &= -(1-P)P^{-2}\left[\frac{(\hat{\lambda}_1^{rd} - \hat{\lambda}_1^r\hat{\lambda}_1^d)}{n_1} + \frac{(\hat{\lambda}_2^{rd} - \hat{\lambda}_2^r\hat{\lambda}_2^d)}{n_2}\right]\end{aligned} \tag{2.25}$$

One may use the estimator

$$\hat{\pi}_{AF}^* = \hat{w}\hat{\pi}_A(1) + (1-\hat{w})\hat{\pi}_A(2)$$

where $\hat{w} = (\hat{\sigma}_{22} - \hat{\sigma}_{12})/(\hat{\sigma}_{11} + \hat{\sigma}_{22} - 2\hat{\sigma}_{12})$. Although the estimator $\hat{\pi}_{AF}^*$ will not be unbiased, it will have nice large-sample properties. In particular, as Folsom *et al.* (1973) point out, $\hat{\pi}_{AF}^*$ will be Neyman's modified chi-square estimator and will belong to the class of best asymptotically normal estimators.

Example 2.1. In a survey to estimate the proportion of habitual gamblers among industrial workers, the model involving two unrelated characters was employed with the following:

Question A: Are you a habitual gambler?
Question Y_1: Do you like the game of cricket?
Question Y_2: Does your date of birth fall in the months Jaunary through June?

Two samples, each of size 50, were taken and each interviewee was required to reply to one randomized and one direct question according to the scheme described at the beginning of this section. In both samples, the sensitive question A was selected with probability $P=0.3$. The observations were as shown in Table 2.2.

TABLE 2.2

Sample 1				Sample 2			
DR \ RR	Yes	No	Total	DR \ RR	Yes	No	Total
Yes	10	14	24	Yes	6	4	10
No	4	22	26	No	17	23	40
Total	14	36	50	Total	23	27	50

Then $n_1=n_2=50$, $P=0.3$, $\hat{\lambda}_1^r=14/50=0.28$, $\hat{\lambda}_1^d=24/50=0.48$, and $\hat{\lambda}_1^{rd}=10/50=0.20$, and similarly, $\hat{\lambda}_2^r=0.46$, $\hat{\lambda}_2^d=0.20$, and $\hat{\lambda}_2^{rd}=0.12$. Hence by (2.20a, b), $\hat{\pi}_A(1)=0.47$ and $\hat{\pi}_A(2)=0.41$. Now, by (2.25), $\hat{\sigma}_{11}=0.0622$, $\hat{\sigma}_{22}=0.0824$, and $\hat{\sigma}_{12}=-0.0146$. Consequently, $\hat{w}=0.56$ and $\hat{\pi}^*_{AF}=0.44$.

In order to compare the model involving two unrelated characters with that involving a single unrelated character, assume that $\pi_{Y_1}=\pi_{Y_2}=\pi_Y$, say, and that Y_1, Y_2, and A are all independent. Then

$$\pi_{Y_1Y_2}=\pi_Y^2 \qquad \pi_{AY_1}=\pi_{AY_2}=\pi_A\pi_Y$$

Now, if $n_1 = n_2 = n/2$, n being the total sample size, it follows from (2.19), (2.23), and (2.24) that

$$\lambda_i^r = P\pi_A + (1-P)\pi_Y \qquad \lambda_i^d = \pi_Y \qquad \lambda_i^{rd} = \pi_Y[P\pi_A + (1-P)\pi_Y] \qquad (i=1,2)$$

$$\sigma_{ii} = 2(nP^2)^{-1}[\lambda_1^r(1-\lambda_1^r) + (1-P)^2\pi_Y(1-\pi_Y)] \qquad (i=1,2)$$

$$\sigma_{12} = 0$$

Hence by (2.21),

$$\text{var}(\hat{\pi}_{AF}) = (nP^2)^{-1}[\lambda_1^r(1-\lambda_1^r) + (1-P)^2\pi_Y(1-\pi_Y)] \tag{2.26}$$

The above can be compared directly with the corresponding variance form (2.10) for Moors' optimized version of the two-sample one-unrelated-question model. Note that λ_1^r and P in (2.26) are equivalent, respectively, to λ_1 and P_1 in (2.10). Comparing (2.26) and (2.10), the relative efficiency of $\hat{\pi}_{AF}$ is

$$1 + \frac{2(1-P)[\lambda_1(1-\lambda_1)\pi_Y(1-\pi_Y)]^{1/2}}{\lambda_1(1-\lambda_1) + (1-P)^2\pi_Y(1-\pi_Y)} \qquad (\geq 1)$$

Clearly, $\hat{\pi}_{AF}$ cannot be less efficient than Moors' estimator under the assumption stated.

In the case of known π_Y, however, a comparison between (2.26) and (2.3), noting that λ_1^r in (2.26) is equivalent to λ in (2.3), shows the relative efficiency of $\hat{\pi}_{AF}$ as

$$\frac{\lambda(1-\lambda)}{\lambda(1-\lambda) + (1-P)^2\pi_Y(1-\pi_Y)} \qquad (\leq 1)$$

implying that the two-alternative-question model can never be more efficient.

2.7. IMPLICIT RANDOMIZATION

Takahasi and Sakasegawa (1977) suggest a version of the unrelated-question technique for estimating the proportion π_A of a dichotomous sensitive character A without the explicit use of a randomizing device. Three independent SRSWRs of sizes n_1, n_2, and n_3 are drawn. Each respondent makes a silent choice, unnoticed by the interviewer, of one item from among three items, say one color from among three colors—violet,

TABLE 2.3

Color \ Attribute	Sample 1		Sample 2		Sample 3	
	A	Ā	A	Ā	A	Ā
Violet	0	1	1	0	1	0
Blue	1	0	0	1	1	0
Green	1	0	1	0	0	1

blue, and green—and reports only 0 or 1, according to the scheme presented in Table 2.3. For example, a respondent in sample 1 who likes the color blue will report 1 or 0 depending on whether he or she belongs to A or Ā, respectively.

Let $\pi_{AV}(\pi_{\bar{A}V})$ be the true proportion of those who belong to A(Ā) and like the violet color. Similarly, considering the colors blue and green, define π_{AB}, $\pi_{\bar{A}B}$, π_{AG}, and $\pi_{\bar{A}G}$. Note that

$$\pi_A = \pi_{AV} + \pi_{AB} + \pi_{AG} \qquad 1 - \pi_A = \pi_{\bar{A}V} + \pi_{\bar{A}B} + \pi_{\bar{A}G} \tag{2.27}$$

For $i = 1, 2, 3$, denoting by λ_i the probability of the response 1 in the ith sample, it is then obvious that

$$\lambda_1 = \pi_{AB} + \pi_{AG} + \pi_{\bar{A}V} \qquad \lambda_2 = \pi_{AV} + \pi_{AG} + \pi_{\bar{A}B}$$
$$\lambda_3 = \pi_{AV} + \pi_{AB} + \pi_{\bar{A}G}$$

By (2.27),

$$\lambda_1 + \lambda_2 + \lambda_3 = \pi_A + 1$$

Hence, if $\hat{\lambda}_i$ is the sample proportion corresponding to λ_i, a UE of π_A is obtained as

$$\hat{\pi}_{ATS} = \hat{\lambda}_1 + \hat{\lambda}_2 + \hat{\lambda}_3 - 1 \tag{2.28}$$

It is a simple exercise (Exercise 2.3) to find the variance of $\hat{\pi}_{ATS}$ and a UE of this variance.

EXERCISES

2.1. (Moors, 1971) Show that if $\pi_Y = 1/2$, then (2.15) reduces to

$$\text{var}(\hat{\pi}_{AU2}) = \frac{\pi_A(1-\pi_A)}{n} + \frac{1-P}{nP^2}\left[\frac{1}{2} + \sqrt{\lambda_1(1-\lambda_1)}\right]$$

where $\lambda_1 = P\pi_A + (1/2)(1-P)$. Observe that

$$\lambda_1(1-\lambda_1) = \frac{1}{4} - P^2\left(\pi_A - \frac{1}{2}\right)^2 \leq \frac{1}{4}$$

Hence, from (1.3), conclude that

$$\text{var}(\hat{\pi}_{AU2}) < \text{var}(\hat{\pi}_{AW})$$

provided that $P > 1/2$. This proves Theorem 2.2 in the special case $\pi_Y = 1/2$, $P > 1/2$. [Hint: For the last part, use the fact that for $P > 1/2$, $0 < (2P-1) < P^2$, and hence $(2P-1)^2 < P^4 < P^3$.]

2.2. (Folsom, *et al.*, 1973) Verify (2.23c) and (2.24) explicitly. [Hint: For $i = 1, 2$, $\hat{\lambda}_i^r = \hat{\lambda}_i^{rd} + (\hat{\lambda}_i^r - \hat{\lambda}_i^{rd})$, $\hat{\lambda}_i^d = \hat{\lambda}_i^{rd} + (\hat{\lambda}_i^d - \hat{\lambda}_i^{rd})$, and $\hat{\lambda}_i^{rd}$, $\hat{\lambda}_i^r - \hat{\lambda}_i^{rd}$, $\hat{\lambda}_i^d - \hat{\lambda}_i^{rd}$ are multinomial proportions relating to disjoint classes, the corresponding true proportions being λ_i^{rd}, $\lambda_i^r - \lambda_i^{rd}$ and $\lambda_i^d - \lambda_i^{rd}$.]

2.3. (Takahasi and Sakasegawa, 1977) From (2.28), derive $\text{var}(\hat{\pi}_{ATS})$ and suggest a UE of this variance. Also show that

$$\frac{3}{n}\pi_A(1-\pi_A) \leq \text{var}(\hat{\pi}_{ATS}) \leq \frac{1}{n}[2 + \pi_A(1-\pi_A)]$$

2.4. (Takahasi and Sakasegawa, 1977) There can be several schemes for implicit randomization and the one described in Section 2.7 is, by no means, unique. In the setup of Section 2.7, suppose that the interviewees respond according to the plan presented in Table 2.4. Show that (2.28) still represents a UE of π_A, although Table 2.4 is quite different from Table 2.3.

2.5. (Bourke, 1974) With a single unrelated character, consider a design that uses a deck of cards made up of three types of cards, as in Figure 2.1. In the figure $A(\bar{A})$ and $Y(\bar{Y})$ are abbreviations for the statements "I belong to group $A(\bar{A})$" and "I belong to group $Y(\bar{Y})$," respectively. With an SRSWR of size n, each respondent randomly selects a card from the deck and describes his or her true character with a response (i, j), $i, j = 1, 2$, considering the upper and lower halves of the card. For example, if a respondent belongs to A and $\bar{Y}$ and

TABLE 2.4

Color \ Attribute	Sample 1 A	Sample 1 $\bar{A}$	Sample 2 A	Sample 2 $\bar{A}$	Sample 3 A	Sample 3 $\bar{A}$
Violet	1	0	0	0	1	1
Blue	1	1	1	0	0	0
Green	0	0	1	1	1	0

draws a card of type 3, his or her response is (2, 2). Let P_1, P_2, and P_3 be the proportions ($P_1+P_2+P_3=1$) of card types 1, 2, and 3 in the deck, and let λ_{ij} be the probability of the response (i, j). Show that

$$\lambda_{11}=P_1\pi_A\pi_Y+P_3\pi_Y(1-\pi_A)$$
$$\lambda_{12}=P_1\pi_A(1-\pi_Y)+P_2(1-\pi_A)+P_3\pi_A\pi_Y$$
$$\lambda_{21}=P_1(1-\pi_A)\pi_Y+P_2\pi_A+P_3(1-\pi_A)(1-\pi_Y)$$
$$\lambda_{22}=P_1(1-\pi_A)(1-\pi_Y)+P_3\pi_A(1-\pi_Y)$$

provided that the sensitive and unrelated characters are independent. Observe that

$$\lambda_{1\cdot}=\lambda_{11}+\lambda_{12}=(P_1-P_2)\pi_A+P_3\pi_Y+P_2$$
$$\lambda_{\cdot 1}=\lambda_{11}+\lambda_{21}=P_1\pi_Y+(P_2-P_3)\pi_A+P_3$$

and hence

$$\pi_A=D^{-1}[P_1(\lambda_{1\cdot}-P_2)-P_3(\lambda_{\cdot 1}-P_3)]$$

Card type 2	Card type 3
(1) $\bar{A}$ (2) A	(1) Y (2) $\bar{Y}$
(1) A (2) $\bar{A}$	(1) $\bar{A}$ (2) A

FIGURE 2.1

provided that $D = P_1(P_1 - P_2) - P_3(P_2 - P_3) \neq 0$. Assuming that $D \neq 0$, suggest a UE of π_A in terms of the sample proportions corresponding to the λ_{ij}'s. Obtain the variance of the UE and examine how this variance can be unbiasedly estimated.

REFERENCES

Boruch, R. F. (1972). Relations among statistical methods for assuring confidentiality of social research data. *Soc. Sci. Res.* **1**, 403–414.

Bourke, P. D. (1974). Vector response in RR designs. Private communication.

Dowling, T. A., and Shachtman, R. (1975). On the relative efficiency of RR models. *J. Amer. Statist. Assoc.* **70**, 84–87.

Folsom, R. E., Greenberg, B. G., Horvitz, D. G., and Abernathy, J. R. (1973). The two alternate questions RR model for human surveys. *J. Amer. Statist. Assoc.* **68**, 525–530.

Greenberg, B. G., Abul-Ela, Abdel-Latif, A., Simmons, W. R., and Horvitz, D. G. (1969). The unrelated question RR model: theoretical framework. *J. Amer. Statist. Assoc.* **64**, 520–539.

Horvitz, D. G., Shah, B. V., and Simmons, W. R. (1967). The unrelated question RR model. *Proc. ASA Soc. Statist. Sec.*, 65–72.

Horvitz, D. G., Greenberg, B. G., and Abernathy, J. R. (1976) RR: a data gathering device for sensitive questions. *Internat. Statist. Rev.* **44**, 181–196.

Lanke, J. (1975). On the choice of unrelated question in Simmons' version of RR. *J. Amer. Statist. Assoc.* **70**, 80–83.

Moors, J. J. A. (1971). Optimization of the unrelated question RR model. *J. Amer. Statist. Assoc.* **66**, 627–629.

Takahasi, K. and Sakasegawa, H. (1977). An RR technique without use of any randomizing device. *Ann. Inst. Statist. Math.* **29**, 1–8.

Appendix

Supplementary Remarks on the Unrelated-Question Model

A2.1 UNBIASED AND MAXIMUM LIKELIHOOD ESTIMATION

Referring to relation (2.1), we may observe that although $\pi_A \in [0, 1]$, because of the parameter $P(\neq 0, 1)$ being fixed, as also π_Y being fixed in the left-open, right-closed interval (0, 1], we have

$$(1-P)\pi_Y < \lambda < P + (1-P)\pi_Y \tag{A2.1}$$

In other words, although λ is a probability (that of a "yes" reply in Simmons' RR device), it is not permitted a value in the entire interval [0, 1], only within a subinterval as specified in (A2.1). Treating this as a problem of estimation with a truncated parametric space, Moors (1985) argued that λ (and hence π_A) does not admit a UE for the reasons given in Section A1.2. Of course, with $\hat{\lambda} = n_1/n$ as the proportion of "yes" replies in an SRSWR of size n and given P and π_Y,

$$\hat{\pi}_{AU1} = \frac{(\hat{\lambda} - (1-P)\,\pi_Y)}{P}$$

may take values outside [0, 1] and hence not admitted by Moors (1985) as an estimator. But if one ignores his warning, it continues to be the unique UE for π_A in terms of the survey data available as "yes-no" replies from sampled persons. Also, when π_Y is unknown and two SRSWORs of sizes n_i are chosen, using P_i $(i=1, 2)$, $P_1 \neq P_2$, as RR design parameters described in Section 2.3, yielding sample proportions $\hat{\lambda}_i = n_{i1}/n_i$, $i=1, 2$, one has the unique UE for π_A as

$$\hat{\pi}_{AU2} = \frac{\hat{\lambda}_1(1-P_2)-\hat{\lambda}_2(1-P_1)}{P_1-P_2}$$

in terms of survey data available through "yes-no" responses from sampled persons. But contrary to the claim of Greenberg *et al.* (1969), neither $\hat{\pi}_{AU1}$ nor $\hat{\pi}_{AU2}$ is an MLE for π_A based on the respective survey data described above.

Flinger *et al.* (1977) have shown that the MLE for π_A, when π_Y is known and a single sample is drawn and used, is $\tilde{\pi}_{AU1}$ given by

$$\begin{aligned} \tilde{\pi}_{AU1} &= \hat{\pi}_{AU1} && \text{if } (1-P)\pi_Y < \hat{\lambda} < P + (1-P)\pi_Y \\ &= 1 && \text{if } \hat{\lambda} \geq P + (1-P)\pi_Y \\ &= 0 && \text{if } \hat{\lambda} < (1-P)\pi_Y \end{aligned}$$

This MLE is clearly a biased estimator for π_A and, as in Section A1.3, MSE $(\tilde{\pi}_{AU1}) < \text{var}(\hat{\pi}_{AU1})$.

A2.2 SRSWOR WITH SIMMONS' RRT

Let π_Y be unknown and two independent samples be drawn in n_i $(i=1, 2)$ draws from a finite population of N units and for the ith sample a selected person answer with probability P_i about bearing the sensitive character A and with $(1-P_i)$ about bearing the innocuous character Y, $P_1 \neq P_2$. For the ith sample let M_i cards of two types in proportions $P_i/(1-P_i)$ be used to choose from $(i=1,2)$. Except for the changes in n_i, M_i, P_i $(i=1,2)$ we will suppose the two samples to be chosen and the RR device to be employed in an identical and independent manner.

For schemes I to IV as in Section A1.4, one may note from Kim and Flueck (1978) that as in Section A2.1, $\hat{\pi}_{AU2}$, is unbiased for π_A. Its variances under the respective schemes may be compared along the lines of Section A1.4.

A2.3 SYMMETRY OF RESPONSE

Consider the Simmons model with π_Y known. Here a "yes" ("no") reply implies that a respondent has (does not have) either the character A or Y. Negation of A and Y will, as usual, be denoted by $\bar{A}$ and $\bar{Y}$. But with the

Warner model a "yes" ("no") implies bearing either A or its complement $\bar{A}$. So the Warner model has a symmetry of response, whereas the Simmons model does not. Therefore, Bourke (1974a, b) suggests modifications in the RR device to render symmetry to the latter. A way to do so is to apply two-way randomization. By way of illustration, let us give an example due to Bourke.

Suppose that cards of type I and type II in proportions $P/(1-P)$ are used. Each card of either type has two sections, I and II. Once a card is chosen at random its sections I and II are chosen with probabilities Q and $(1-Q)$, respectively. If for a type I card, section I is chosen, a respondent is to report 1(2) if he or she bears $A(\bar{A})$ and if section II is chosen, the truthful report would be 1(2) if he or she bears $Y(\bar{Y})$. Similarly, for a type II card, if section I is chosen, a truthful report would be 1(2) if one bears $\bar{Y}(Y)$ and if section II is chosen, a correct report would be 1(2) if one bears $\bar{A}(A)$, respectively.

Based on an SRSWR of size n, the survey data will produce n vector responses (i, j), $i, j = 1, 2$, to yield a summary vector response (m_1, m_2) with $m_1(m_2)$ equal to the frequency of 1(2). Assuming A and Y as independent characters, it is not difficult to obtain an unbiased estimator for π_A, provided that π_Y be assumed known.

This scheme apparently ensures symmetry of response and may be expected to instill in a respondent a sense that his or her privacy is being maintained.

EXERCISES

A1. Derive the MLE corresponding to $\hat{\pi}_{AU2}$. Show that the MSE of this MLE is less than var $(\hat{\pi}_{AU2})$.

A2. Find the MLE corresponding to $\hat{\pi}_{ATS}$ in (2.28). Show that the MSE of this MLE is smaller than var $(\hat{\pi}_{ATS})$.

A3. (a) Obtain a UE for π_A for the RR scheme of Bourke in Section A2.3 and find a formula for its variance.
(b) Obtain the MLE for π_A in this case.
(c) Show how you may reduce this RR scheme to Warner's or Simmons' scheme.
(d) Suggest two alternative RR schemes using two-stage randomization and ensuring symmetry of response. Pose and answer problems as in parts (a) to (c).

REFERENCES

Bourke, P. D. (1974a). Symmetry of response in RR designs. *Tech. Rep.* 75. Institute of Statistics, University of Stockholm, Sweden.

Bourke, P. D. (1974b). Vector response in RR designs. Private Communication.

Flinger, M. A., Policello, G. E., and Singh, J. (1977). A comparison of two RR survey methods with consideration for the level of respondent protection. *Comm. Statist—Theory Methods* **6**, 1511–1524.

Greenberg, B. G., Abul-Ela, Abdel-Latif, A., Simmons, W. R., and Horvitz, D. G. (1969). The unrelated question RR model: theoretical framework. *J. Amer. Statist. Assoc.* **64**, 520–539.

Kim, Jong-Ik, and Flueck, John A. (1978). Modifications of the randomized response technique for sampling without replacement. *Proc. ASA Sec. Surv. Res. Methods.* 346–350.

Moors, J. J. A. (1985). Estimation in truncated parameter spaces. Ph.D. thesis, Katholieke Hogeschool, Tilburg.

3

Polychotomous Population and Multiattribute Situations

3.1. INTRODUCTION

The RR techniques presented so far deal with a single dichotomous sensitive attribute. As one can imagine, the methods can be extended in several directions. First, in many practical situations, there may be more than two naturally occurring classes according to the character under consideration. For example (see Abul-Ela *et al.*, 1967), in a survey to estimate the proportion of unmarried mothers in a region, one may think of three classes, consisting of (1) females already married upon pregnancy, (2) females pregnant at the time of marriage, and (3) females unmarried at the time of childbirth.

Second, in some investigations several stigmatizing characters have to be considered together and the objective may be to study their possible interrelationship. For example, interest may lie in examining the extent to which alcoholism is linked with gambling. In this chapter we present extensions of the Warner and unrelated-question models to cover such polychotomous and/or multiattribute situations.

3.2. SOME TECHNIQUES FOR A POLYCHOTOMOUS POPULATION

To begin with, consider a single sensitive character according to which there are t (≥ 2) classes. Thus the population is now classified into t mutually exclusive and exhaustive groups at least one and at most $t-1$ of which are stigmatizing. The problem is to estimate the corresponding true proportions $\pi_1, \ldots, \pi_t$, where $0<\pi_j<1$ ($j=1, \ldots, t$) and $\sum_{j=1}^{t} \pi_j=1$.

Abul-Ela *et al.* (1967) extend the Warner technique to polychotomous populations. In their method, $s\ (=t-1)$ SRSWRs of sizes $n_1, \ldots, n_s$ are drawn independently from the population. As usual, a random device is needed to collect data on the composition of the groups without asking the respondent to reveal his or her own group. The device consists of s decks of cards, deck i for the ith sample $(i=1, \ldots, s)$. Each deck has t different types of cards, the jth type announcing a membership in the jth group $(j=1, \ldots, t)$. A sampled respondent is asked to draw one card at random from the deck specified for the particular sample he or she belongs to and to read the statement on the card unobserved by the interviewer. If the statement specifies the correct group of the interviewee, he or she reports "yes," otherwise "no."

For $i=1, \ldots, s; j=1, \ldots, t$, let P_{ij} denote the porportion of cards with the statement expressing membership in the jth group for the ith sample. Then $\Sigma_{j=1}^{t} P_{ij}=1$ and the probability of a "yes" response for any interviewee in the ith sample is, say.

$$\lambda_i=\sum_{j=1}^{t} P_{ij}\,\pi_j \qquad (i=1, \ldots, s)$$

Since $s=t-1$ and $\Sigma_{j=1}^{t} \pi_j=1$, it follows that

$$\sum_{j=1}^{s} (P_{ij}-P_{it})\pi_j=\lambda_i-P_{it} \qquad (i=1, \ldots, s)$$

or in matrix notation,

$$P\boldsymbol{\pi}=\boldsymbol{\xi} \tag{3.1}$$

where

$$P=\begin{bmatrix} P_{11}-P_{1t} & P_{12}-P_{1t} & \cdots & P_{1s}-P_{1t} \\ P_{21}-P_{2t} & P_{22}-P_{2t} & \cdots & P_{2s}-P_{2t} \\ & \vdots & & \\ P_{s1}-P_{st} & P_{s2}-P_{st} & \cdots & P_{ss}-P_{st} \end{bmatrix}$$

$\boldsymbol{\pi}=(\pi_1, \ldots, \pi_s)', \boldsymbol{\xi}=(\lambda_1-P_{1t}, \ldots, \lambda_s-P_{st})'.$

If in the ith sample n_{i1} persons report "yes," then a UE of λ_i is $\hat{\lambda}_i=n_{i1}/n_i$ $(i=1, \ldots, s)$ and hence writing

$$\mathbf{c}=(\hat{\lambda}_1-P_{1t}, \ldots, \hat{\lambda}_s-P_{st})'$$

from (3.1), a UE of $\boldsymbol{\pi}$ is obtained as

$$\hat{\boldsymbol{\pi}}=(\hat{\pi}_1, \ldots, \hat{\pi}_s)'=P^{-1}\mathbf{c}, \tag{3.2}$$

provided that P is nonsingular. Clearly, then, π_t may be estimated unbiasedly by $\hat{\pi}_t = 1 - \Sigma_{j=1}^{s} \hat{\pi}_j$.

Since the s samples are drawn independently, $n_{11}, \ldots, n_{s1}$ and hence $\hat{\lambda}_1, \ldots, \hat{\lambda}_s$ are independent. Also, for each i, n_{i1} has the binominal distribution with parameters n_i and λ_i. Hence the dispersion matrix of $\underline{c}$ is

$$\text{disp}(\mathbf{c}) = \text{diag}(V_{11}, \ldots, V_{ss})$$

where $V_{ii} = \lambda_i(1-\lambda_i)/n_i$. By (3.2), it follows that the dispersion matrix of $\hat{\boldsymbol{\pi}}$ is given by

$$\text{disp}(\hat{\boldsymbol{\pi}}) = P^{-1}\,\text{diag}(V_{11}, \ldots, V_{ss})\,(P^{-1})' \tag{3.3}$$

For each i, $\hat{V}_{ii} = \hat{\lambda}_i(1-\hat{\lambda}_i)/(n_i - 1)$ is a UE of V_{ii}. Hence disp($\hat{\boldsymbol{\pi}}$) may be unbiasedly estimated by

$$\widehat{\text{disp}}(\hat{\boldsymbol{\pi}}) = P^{-1}\,\text{diag}(\hat{V}_{11}, \ldots, \hat{V}_{ss})\,(P^{-1})' \tag{3.4}$$

The observations following (1.4) in Chapter 1 extend themselves to the present setting. Thus var($\hat{\pi}_j$)'s, as functions of P_{ij}'s decrease rapidly as P_{ij}'s depart from equality (i.e., from $1/t$). But if one chooses a very small value for some P_{ij}, the cards representing that proportion might not easily be observed by the respondents. This will lessen the degree of cooperation expected from them.

Example 3.1. In an investigation on the prevalence of pregnancy before marriage, the objective was to estimate the proportions π_1, π_2, and π_3 in three mutually exclusive and exhaustive classes consisting of

1. Females married at the time they became pregnant.
2. Females married during pregnancy.
3. Females unmarried at the time of childbirth.

Clearly, group (1) is innocuous and group (3) is the most stigmatizing.

Two SRSWRs, consisting of 180 and 200 females, were drawn independently. Each respondent in the first (second) sample was supplied with a deck of 100 cards, of which 40, 30, and 30 (30, 30, and 40) announced memberships in groups 1, 2, and 3, respectively. Each person sampled was required to draw one card after shuffling the deck, unobserved by the interviewer, and report only "yes" or "no" according to whether or not the statement on the card drawn specified her correct group. There were 70 and 61 "yes" responses from the two samples.

In order to analyze the data, note that here $t=3$, $s=2$, $n_1=180$, $n_2=200$, $P_{11}=0.4$, $P_{12}=P_{13}=0.3$, $P_{21}=P_{22}=0.3$, and $P_{23}=0.4$. Hence

$$P = \begin{bmatrix} 0.1 & 0 \\ -0.1 & -0.1 \end{bmatrix}$$

which is nonsingular with

$$P^{-1} = \begin{bmatrix} 10 & 0 \\ -10 & -10 \end{bmatrix}$$

Also, $n_{11} = 70$, $n_{21} = 61$, $\hat{\lambda}_1 = 0.3889$, $\hat{\lambda}_2 = 0.3050$, and $\underline{c} = (0.0889, -0.0950)'$. Hence, by (3.2), an unbiased estimate of $\pi = (\pi_1, \pi_2)'$ is obtained as

$$\hat{\pi} = (0.889, 0.061)'$$

Thus $\hat{\pi}_1 = 0.889$, $\hat{\pi}_2 = 0.061$, and $\hat{\pi}_3 = 1 - \hat{\pi}_1 - \hat{\pi}_2 = 0.050$ are the estimated proportions in the three groups. Since $\hat{V}_{11} = 1.3276 \times 10^{-3}$, $\hat{V}_{22} = 1.0652 \times 10^{-3}$, from (3.4),

$$\hat{\text{disp}}(\hat{\pi}) = \begin{bmatrix} 0.133 & -0.133 \\ -0.133 & 0.239 \end{bmatrix}$$

Hence $\hat{\text{var}}(\hat{\pi}_1) = 0.133$ and $\hat{\text{var}}(\hat{\pi}_2) = 0.239$. Also,

$$\begin{aligned} \hat{\text{var}}(\hat{\pi}_3) &= \hat{\text{var}}(1 - \hat{\pi}_1 - \hat{\pi}_2) \\ &= \hat{\text{var}}(\hat{\pi}_1) + \hat{\text{var}}(\hat{\pi}_2) + 2\,\hat{\text{cov}}(\hat{\pi}_1, \hat{\pi}_2) \\ &= 0.106 \end{aligned}$$

For polychotomous populations with t groups, Eriksson (1973) suggests two alternative approaches. As before, from the population $s\,(=t-1)$ independent SRSWR's of sizes $n_1, \ldots, n_s$ are drawn. In the first approach, each interviewee in the ith sample encounters a statement announcing membership in group i (with probability P_i) or a statement announcing membership in group t (with probability $1 - P_i$), so that the probability of a "yes" response in the ith sample is

$$\lambda_i = P_i \pi_i + (1 - P_i)\pi_t \qquad (i = 1, \ldots, s)$$

Since $s = t - 1$ and $\sum_{j=1}^{t} \pi_j = 1$, it follows as before that

$$(2P_i - 1)\pi_i - (1 - P_i) \sum_{\substack{j=1 \\ j \neq i}}^{s} \pi_j = \lambda_i - (1 - P_i) \qquad (i = 1, \ldots, s)$$

or, analogously to (3.1), in matrix notation,

$$P\pi = \xi$$

where now

$$P=\begin{bmatrix} 2P_1-1 & -(1-P_1) & \cdots & -(1-P_1) \\ -(1-P_2) & 2P_2-1 & \cdots & -(1-P_2) \\ & \vdots & & \\ -(1-P_s) & -(1-P_s) & \cdots & 2P_s-1 \end{bmatrix}$$

$\pi=(\pi_1, \ldots, \pi_s)'$, and $\xi=(\lambda_1-1+P_1, \ldots, \lambda_s-1+P_s)'$.

If n_{i1} denotes the number of "yes" responses in the ith sample, defining $\hat{\lambda}_i=n_{i1}/n_i$ $(i=1, \ldots, s)$ and

$$\mathbf{c}=(\hat{\lambda}_1-1+P_1, \ldots, \hat{\lambda}_s-1+P_s)'$$

a UE of π follows as $\hat{\pi}=P^{-1}\mathbf{c}$, provided that P is nonsingular. The dispersion matrix of $\hat{\pi}$ and a UE of this dispersion matrix may be obtained exactly along the lines of (3.3) and (3.4).

In the second approach of Eriksson (1973), each interviewee in the ith sample encounters a statement declaring membership in group i (with probability P_i) or a statement declining membership in group i (with probability $1-P_i$), $i=1, \ldots, s$. As before, the probability of a "yes" response in the ith sample is

$$\lambda_i=P_i\pi_i+(1-P_i)(1-\pi_i) \qquad (i=1, \ldots, s). \tag{3.5}$$

Hence recalling that $\Sigma_{j=1}^{t}\pi_j=1$, it is possible as before to obtain UEs of $\pi_1, \ldots, \pi_t$, provided that a suitable nonsingularity condition holds. In Exercise 3.1 we ask the reader to examine the details of this approach.

Whereas the techniques described above are essentially different types of extensions of the Warner model to a polychotomous setup, Greenberg *et al.* (1969) investigate an extension of the unrelated-question model to such a situation. Consider as before a population classified into t groups according to a sensitive character. Let Y be an unrelated dichotomous character and π_Y be the unknown true proportion of Y. With t independent SRSWRs of sizes $n_1, \ldots, n_t$, suppose the randomization device is such that each respondent in the ith sample encounters a statement announcing membership in group j (with probability P_{ij}), $j=1, \ldots, t-1$, or a statement announcing membership in group Y (with probability P_{it}). Here $\Sigma_{j=1}^{t} P_{ij}=1$ and the probability of a "yes" response from any interviewee in the ith sample is

$$\lambda_i=\sum_{j=1}^{t-1} P_{ij}\pi_j+P_{it}\pi_Y \qquad (i=1, \ldots, t)$$

Defining $\boldsymbol{\lambda}=(\lambda_1, \ldots, \lambda_t)'$, $\boldsymbol{\pi}^*=(\pi_1, \ldots, \pi_{t-1}, \pi_Y)'$, and $P=((P_{ij}))$, one obtains $\boldsymbol{\lambda}=P\pi^*$. Hence if $\hat{\lambda}_1, \ldots, \hat{\lambda}_t$ are the sample proportions corresponding to $\lambda_1, \ldots, \lambda_t$, and $\hat{\lambda}=(\hat{\lambda}_1, \ldots, \hat{\lambda}_t)'$, a UE of $\boldsymbol{\pi}^*$ emerges as before as

$$\hat{\boldsymbol{\pi}}^*=(\hat{\pi}_1, \ldots, \hat{\pi}_{t-1}, \hat{\pi}_Y)=P^{-1}\hat{\lambda} \tag{3.6}$$

provided that P is nonsingular. An obvious UE of π_t is then $\hat{\pi}_t=1-\Sigma_{j=1}^{t-1}\hat{\pi}_j$. As in (3.3) and (3.4), the dispersion matrix of $\hat{\boldsymbol{\pi}}^*$ and a UE thereof are given by

$$\begin{aligned} \operatorname{disp}(\hat{\boldsymbol{\pi}}^*)&=P^{-1}\operatorname{diag}(V_{11}, \ldots, V_{tt})(P^{-1})', \\ \widehat{\operatorname{disp}}(\hat{\boldsymbol{\pi}}^*)&=P^{-1}\operatorname{diag}(\hat{V}_{11}, \ldots, \hat{V}_{tt})(P^{-1})', \end{aligned} \tag{3.7}$$

where $V_{ii}=\lambda_i(1-\lambda_i)/n_i$, $\hat{V}_{ii}=\hat{\lambda}_i(1-\hat{\lambda}_i)/(n_i-1)$, $i=1, \ldots, t$.

Example 3.2. In an election, there were three candidates and a survey was conducted, before the actual polling, to estimate the proportions π_1, π_2, and π_3, supporting them. The unrelated-question model for a polychotomous population was adopted as follows.

Three SRSWRs of sizes 100, 150, and 200 were drawn independently. Each interviewee was provided with a deck of 80 cards. There were three different types of cards in a deck, the card types and the corresponding frequencies being as shown in Table 3.1. Observe that the question on cards of the third type was unrelated to the character under study, namely voting preference. Each interviewee was asked to draw one card from the deck supplied, unobserved by the interviewer as usual, and reply to the question on the card drawn. There were 37, 60, and 72 "yes" responses in the three samples, respectively.

Turning to the analysis of the data, observe that here $t=3$, $n_1=100$, $n_2=150$, $n_3=200$, and

$$P=\begin{bmatrix} 0.375 & 0.375 & 0.250 \\ 0.375 & 0.250 & 0.375 \\ 0.250 & 0.375 & 0.375 \end{bmatrix}$$

TABLE 3.1

Card type	Question on the card	Frequency in the deck used for: Sample 1	Sample 2	Sample 3
1	Do you support candidate 1?	30	30	20
2	Do you support candidate 2?	30	20	30
3	Do you prefer tea to coffee?	20	30	30

The matrix P is nonsingular with

$$P^{-1} = \begin{bmatrix} 3 & 3 & -5 \\ 3 & -5 & 3 \\ -5 & 3 & 3 \end{bmatrix}$$

Since $\hat{\lambda}_1 = 0.37$, $\hat{\lambda}_2 = 0.40$, and $\hat{\lambda}_3 = 0.36$, by (3.6),

$$\hat{\boldsymbol{\pi}}^* = (0.51, 0.19, 0.43)'$$

In particular, $\hat{\pi}_1 = 0.51$, $\hat{\pi}_2 = 0.19$, and hence $\hat{\pi}_3 = 0.30$. Also, $\hat{V}_{11} = 2.3545 \times 10^{-3}$, $\hat{V}_{22} = 1.6107 \times 10^{-3}$, and $\hat{V}_{33} = 1.1577 \times 10^{-3}$. Hence, by (3.7),

$$\widehat{\text{disp}}\,(\hat{\boldsymbol{\pi}}^*) = \begin{bmatrix} 0.0646 & -0.0203 & -0.0382 \\ -0.0203 & 0.0719 & -0.0491 \\ -0.0382 & -0.0491 & 0.0838 \end{bmatrix}$$

Evidently, then, $\widehat{\text{var}}\,(\hat{\pi}_1) = 0.0646$, $\widehat{\text{var}}(\hat{\pi}_2) = 0.0719$, $\widehat{\text{cov}}(\hat{\pi}_1, \hat{\pi}_2) = -0.0203$, and as in Example 3.1, $\widehat{\text{var}}\,(\hat{\pi}_3) = 0.0959$.

3.3. USE OF VECTOR RESPONSE

The techniques described in Section 3.2 utilize dichotomous response for dealing with polychotomous populations. It is intuitively clear that if polychotomous (or vector) responses are allowed in such situations, the data acquired through a single sample may be adequate for the unbiased estimation of the proportions in the different classes. This point has been mentioned explicitly by Bourke (1974, 1978). In this section we first develop some general theory for dealing with vector responses and then consider particular applications.

As before, let there be t mutually exclusive and exhaustive classes according to a sensitive attribute, the corresponding unknown proportions being $\pi_1, \ldots \pi_t$. The randomization device is such that an interviewee belonging to the ith category ($i = 1, \ldots t$) reports $1, 2, \ldots$ or t with respective probabilities $p_{1i}, p_{2i}, \ldots p_{ti}$, where

$$\textstyle\sum_{j=1}^{t} p_{ji} = 1 \qquad (i = 1, \ldots, t) \tag{3.8}$$

Then the probability λ_j of the (randomized) response j is given by

$$\lambda_j = \textstyle\sum_{i=1}^{t} p_{ji}\pi_i \qquad (j = 1, \ldots, t)$$

Hence defining $\boldsymbol{\lambda}=(\lambda_1, \ldots, \lambda_t)'$, $\boldsymbol{\pi}=(\pi_1, \ldots, \pi_t)'$, one obtains in matrix notation

$$\boldsymbol{\lambda}=P\boldsymbol{\pi} \tag{3.9}$$

where the square matrix P, of order t, is given by $P=((p_{ji}))$. The matrix P is called the design matrix.

With an SRSWR of size n, let $\hat{\boldsymbol{\lambda}}$ be the vector of sample proportions corresponding to $\boldsymbol{\lambda}$. Then, assuming the nonsingularity of the design matrix P, by (3.9) a UE of $\boldsymbol{\pi}$ emerges as

$$\hat{\boldsymbol{\pi}}=P^{-1}\hat{\boldsymbol{\lambda}} \tag{3.10}$$

Since

$$\text{disp}(\hat{\boldsymbol{\lambda}})=n^{-1}(\lambda^{\delta}-\boldsymbol{\lambda}\boldsymbol{\lambda}')$$

where λ^{δ} is a diagonal matrix with the same diagonal elements as those of $\boldsymbol{\lambda}$ arranged in the same order, one gets

$$\begin{aligned}\text{disp}(\hat{\boldsymbol{\pi}})&=n^{-1}P^{-1}(\lambda^{\delta}-\boldsymbol{\lambda}\boldsymbol{\lambda}')P'^{-1}\\&=n^{-1}(P^{-1}\lambda^{\delta}P'^{-1}-\boldsymbol{\pi}\boldsymbol{\pi}')\end{aligned} \tag{3.11}$$

through an application of (3.9). Clearly, a UE of the dispersion matrix above is given by

$$\widehat{\text{disp}}(\hat{\boldsymbol{\pi}})=(n-1)^{-1}P^{-1}(\hat{\lambda}^{\delta}-\hat{\boldsymbol{\lambda}}\hat{\boldsymbol{\lambda}}')P'^{-1}$$

$\hat{\lambda}^{\delta}$ being a diagonal matrix with the same diagonal elements as those of $\hat{\boldsymbol{\lambda}}$.

A closer examination of the expression for $\text{disp}(\hat{\boldsymbol{\pi}})$ reveals some interesting details. The following lemma (see Mukerjee, 1981) will be helpful in this connection. In what follows, π^{δ} is a diagonal matrix formed from $\boldsymbol{\pi}$ in the same manner as λ^{δ} has been formed from $\boldsymbol{\lambda}$.

Lemma 3.1. The matrix $\lambda^{\delta}-P\pi^{\delta}P'$ is nonnegative definite.

Proof. For any t-component real vector $\underset{\sim}{x}=(x_1, \ldots, x_t)'$, by (3.9),

$$\begin{aligned}\underset{\sim}{x}'(\lambda^{\delta}-P\pi^{\delta}P')\underset{\sim}{x}&=\sum_{j=1}^{t}\lambda_j x_j^2-\sum_{i=1}^{t}\pi_i\left(\sum_{j=1}^{t}p_{ji}x_j\right)^2\\&=\sum_{i=1}^{t}\pi_i\left[\sum_{j=1}^{t}p_{ji}x_j^2-\left(\sum_{j=1}^{t}p_{ji}x_j\right)^2\right]\end{aligned} \tag{3.12}$$

By (3.8) and the Cauchy-Schwarz inequality,

$$\sum_{j=1}^{t}p_{ji}x_j^2\geq\left(\sum_{j=1}^{t}p_{ji}x_j\right)^2 \qquad (i=1, \ldots, t)$$

which shows that (3.12) is nonnegative for every $\underset{\sim}{x}$ and proves the lemma.

By (3.11), disp ($\hat{\boldsymbol{\pi}}$) may be partitioned as

$$\text{disp}(\hat{\boldsymbol{\pi}}) = \Sigma_1 + \Sigma_2 \tag{3.13}$$

where

$$\Sigma_1 = n^{-1}(\pi^\delta - \boldsymbol{\pi}\boldsymbol{\pi}') \qquad \Sigma_2 = n^{-1}P^{-1}(\lambda^\delta - P\pi^\delta P')P'^{-1}$$

Note that Σ_1 is the dispersion matrix of the regular survey estimator of $\boldsymbol{\pi}$, while Σ_2, which is nonnegative definite by Lemma 3.1, represents the component of dispersion associated with the RR experiment. This observation generalizes (1.3) to a polychotomous setup.

Bourke and Dalenius (1976) describe a procedure for actually obtaining the data in a polychotomous situation allowing vector responses. The method is essentially based on an extension of the ideas of Morton (see Section 2.2). The randomization device is such that each respondent states his or her true category (in the form of an integer between 1 and t) with probability p or one of the numbers $1, \ldots, t$ with respective probabilities $p_1, \ldots, p_t$, where $p + \Sigma_{i=1}^t p_i = 1$. Then the probability of getting the (randomized) response j is

$$\begin{aligned} \lambda_j &= p\pi_j + p_j \\ &= (p + p_j)\pi_j + p_j(\pi_1 + \cdots + \pi_{j-1} + \pi_{j+1} + \cdots + \pi_t) \qquad (j = 1, \ldots, t) \end{aligned} \tag{3.14}$$

since $\Sigma_{i=1}^t \pi_i = 1$. Hence the design matrix P becomes

$$P = \begin{bmatrix} p+p_1 & p_1 & \cdots & p_1 \\ p_2 & p+p_2 & \cdots & p_2 \\ & \vdots & & \\ p_t & p_t & \cdots & p+p_t \end{bmatrix}$$

which is nonsingular provided that $p > 0$. On the basis of the sample data one can, of course, employ (3.10) to estimate $\boldsymbol{\pi}$. However, for this particular setting, the relations (3.14) allow a simpler approach that requires no explicit matrix inversion. In fact, defining $\mathbf{p} = (p_1, \ldots, p_t)'$, (3.14) may be expressed as $\boldsymbol{\lambda} = p\boldsymbol{\pi} + \mathbf{p}$ and, as such, a UE of $\boldsymbol{\pi}$ follows as

$$\hat{\boldsymbol{\pi}} = p^{-1}(\hat{\boldsymbol{\lambda}} - \mathbf{p}) \tag{3.15}$$

It is a routine algebra to show that the estimator in (3.15) is precisely the same as that obtained through a direct inversion of the design matrix P and use of (3.10). In Exercise 3.2 we ask the reader to perform the details. It is straightforward to obtain disp($\hat{\boldsymbol{\pi}}$) and a UE of this dispersion matrix.

Liu and Chow (1976) independently describe an elegant randomization device which, incidentally, provides a method of implementing the scheme due to Bourke and Dalenius (1976). The device consists of a jar containing balls of two different colors, say, red and white. On the surface of each white ball one of the numbers 1, 2, . . . , t is marked. The proportion of red balls in the jar is p, while that of white balls marked i is p_i $(i=1, \ldots, t)$ where, obviously, $p+\Sigma_{i=1}^{t} p_i=1$. The jar has a transparent neck where exactly one ball can stand when the jar is turned upside down with a stopper attached to the mouth holding the balls up. Each respondent is asked to turn the jar upside down, shake it thoroughly, and allow a ball to appear in the neck. The person is requested to report his or her true category according to the sensitive character if this ball is red, or the number marked on the ball if it is white. As usual, the experiment is unobserved by the interviewer and the response is one of the numbers 1, . . . , t, the associated probabilities being given by (3.14).

As an illustration, let $t=4$ and suppose that the jar is as in Figure 3.1.

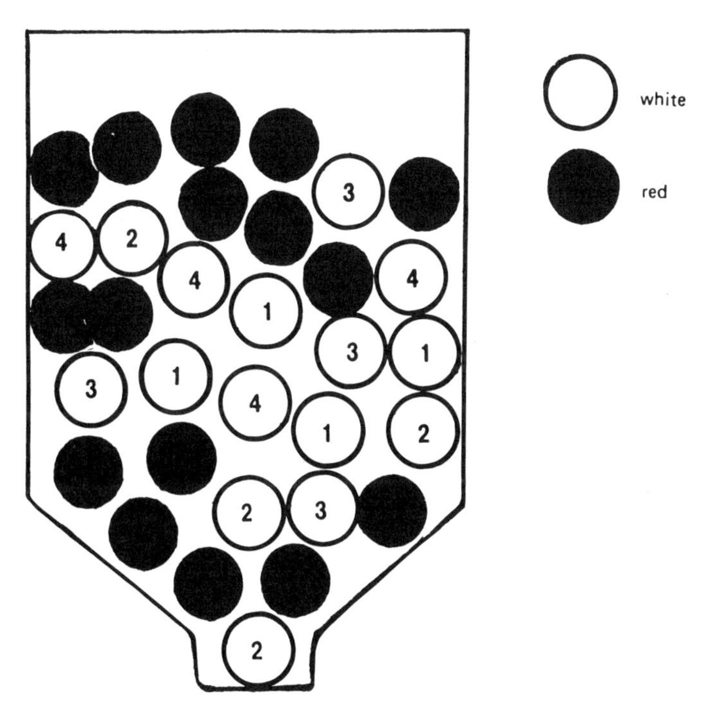

FIGURE 3.1

The jar contains 32 balls, of which 16 are red and 16 white. Each of the numbers 1, 2, 3, 4 is marked on four white balls. Consequently, the values of the design parameters are $p=0.500$, $p_1=p_2=p_3=p_4=0.125$.

Bourke (1974) advocates the application of randomization in two stages, which, he believes, is likely to give a greater feeling of confidentiality to the respondent. One of the schemes suggested by him is described below.

Together with the sensitive character A with t classes, consider an unrelated nonsensitive character U such that there are t classes according to U, the true proportions, namely $\mu_1, \ldots, \mu_t$, for these t classes being known. An SRSWR of size n is drawn. In the first stage of randomization, each respondent draws a card at random from a deck made up of two card types as shown in Figure 3.2. Here S_j and U_j are abbreviations for the statements "I belong to sensitive category j" and "I belong to unrelated category j," respectively. In the second stage, a box of beads, colored blue and red in proportions q_b and $q_r=1-q_b$, is used to choose from which portion of the card the respondent should reply. Let a be the proportion of cards of type 1 in the deck. Then a simple argument based on conditional probabilities shows that the probability λ_j of response j is given by

$$\lambda_j=a(q_b\pi_j+q_r\mu_j)+(1-a)(q_r\pi_j+q_b\mu_j)$$

which is of the form (3.14) with

$$p=aq_b+(1-a)q_r \qquad p_j=[aq_r+(1-a)q_b]\mu_j \qquad (j=1, \ldots, t)$$

Card type 1			Card type 2		
If bead is blue			If bead is blue		
Answer	(1)	S_1	Answer	(1)	U_1
from	(2)	S_2	from	(2)	U_2
here		⋮	here	⋮	
	(t)	S_t		(t)	U_t
If bead is red			If bead is red		
Answer	(1)	U_1	Answer	(1)	S_1
from	(2)	U_2	from	(2)	S_2
here		⋮	here	⋮	
	(t)	U_t		(t)	S_t

FIGURE 3.2

Now (3.15) may be employed for unbiased estimation of $\boldsymbol{\pi}$. In Exercise 3.3 we present another scheme due to Bourke (1974) involving two stages of randomization.

3.4. TECHNIQUES FOR MULTIATTRIBUTE SITUATIONS

The RR techniques for a single sensitive attribute considered in the preceding sections can be extended to multiattribute situations. Eriksson (1973) presents a theory for two-way contingency tables. Drane (1976) considers the problem of testing independence of two sensitive dichotomous attributes, and Mukhopadhyay (1980) deals with RR techniques with two independent sensitive characters. The findings of Bourke (1981) and Mukerjee (1981) are more general and cover the work of Mukhopadhyay (1980). Further results have been obtained by Tamhane (1981). Following Bourke (1981, 1982) and Mukerjee (1981), we first develop a general theory for multiattribute situations.

Let there be m stigmatizing characters $A_1, \ldots, A_m$, the jth character having t_j mutually exclusive and exhaustive categories $A_{j1}, \ldots, A_{jt_j}$. For $i_j = 1, \ldots, t_j; j = 1, \ldots, m$, let $\pi_{i_1 \ldots i_m}$ denote the true proportion corresponding to the category combination $(A_{1i_1}, \ldots, A_{mi_m})$. Let $\boldsymbol{\pi}$ be a vector with elements $\pi_{i_1 \ldots i_m}$, arranged lexicographically. For example, if $m = 3$, $t_1 = t_2 = 2$, $t_3 = 3$, then

$$\boldsymbol{\pi} = (\pi_{111}, \pi_{112}, \pi_{113}, \pi_{121}, \pi_{122}, \pi_{123}, \pi_{211}, \pi_{212}, \pi_{213}, \pi_{221}, \pi_{222}, \pi_{223})'$$

Each respondent in an SRSWR of size n makes m independent RR trials, one for each attribute. The randomization device is such that for the jth attribute ($j = l, \ldots, m$) if the interviewee really belongs to the i_jth category ($i_j = 1, \ldots, t_j$), he or she reports 1, 2, ... or t_j with respective probabilities

$$p_{1i_j}^{(j)}, p_{2i_j}^{(j)}, \ldots, p_{t_j i_j}^{(j)}$$

where

$$\textstyle\sum_{u=1}^{t_j} p_{ui_j}^{(j)} = 1 \tag{3.16}$$

Then if $\lambda_{u_1 \ldots u_m}$ is the probability of getting a response $(u_1, \ldots, u_m)$, considering the m characters together, one obtains

$$\lambda_{u_1 \ldots u_m} = \sum_{i_1=1}^{t_1} \cdots \sum_{i_m=1}^{t_m} \left(\prod_{j=1}^{m} p_{u_j i_j}^{(j)} \right) \pi_{i_1 \ldots i_m}$$

Denoting by $\boldsymbol{\lambda}$ the vector with elements $\lambda_{u_1 \ldots u_m}$, arranged lexicographically, the above can be expressed in matrix notation as

$$\boldsymbol{\lambda}=(P_1 \times \ldots \times P_m)\boldsymbol{\pi} \tag{3.17}$$

where $\times$ stands for the Kronecker product, and for $j=1, \ldots, m$ the square matrix P_j of order t_j is given by

$$P_j=\left(\left(p^{(j)}_{u_j i_j}\right)\right)$$

The matrices P_j are called design matrices. It may be noted that (3.17) is the natural extension of (3.9) when one considers several attributes.

Considering first the problem of estimation of $\boldsymbol{\pi}$, denote by $\hat{\boldsymbol{\lambda}}$ the vector of sample proportions corresponding to $\boldsymbol{\lambda}$ and assume the nonsingularity of the design matrices. Then as

$$(P_1 \times \cdots \times P_m)^{-1}=P_1^{-1} \times \cdots \times P_m^{-1}$$

by (3.17) and analogously to (3.10), a UE of $\boldsymbol{\pi}$ follows as

$$\hat{\boldsymbol{\pi}}=(P_1^{-1} \times \cdots \times P_m^{-1})\hat{\boldsymbol{\lambda}} \tag{3.18}$$

Let $P=P_1 \times \cdots \times P_m$ (so that $P^{-1}=P_1^{-1} \times \cdots \times P_m^{-1}$). Then just as in (3.11),

$$\operatorname{disp}(\hat{\boldsymbol{\pi}})=n^{-1}(P^{-1}\lambda^{\delta}P'^{-1}-\boldsymbol{\pi}\boldsymbol{\pi}') \tag{3.19}$$

while a UE of this dispersion matrix is

$$\begin{aligned}\widehat{\operatorname{disp}}(\hat{\boldsymbol{\pi}})&=(n-1)^{-1}P^{-1}(\hat{\lambda}^{\delta}-\hat{\boldsymbol{\lambda}}\hat{\boldsymbol{\lambda}}')P'^{-1}\\&=(n-1)^{-1}(P^{-1}\hat{\lambda}^{\delta}P'^{-1}-\hat{\boldsymbol{\pi}}\hat{\boldsymbol{\pi}}')\end{aligned} \tag{3.20}$$

by (3.18), λ^{δ} and $\hat{\lambda}^{\delta}$ being as in Section 3.3. It is possible (Exercise 3.4) to decompose $\operatorname{disp}(\hat{\boldsymbol{\pi}})$ into two nonnegative definite components along the line of (3.13) and with similar interpretations.

In a multiattribute setup, one is often interested in examining whether or not the attributes are associated. To that effect, consider the problem of testing the null hypothesis of complete independence of the sensitive characters $A_1, \ldots, A_m$, that is,

$$H_0\colon \boldsymbol{\pi}=\boldsymbol{\pi}^{(1)} \times \cdots \times \boldsymbol{\pi}^{(m)}$$

where for $j=1, \ldots, m$, $\boldsymbol{\pi}^{(j)}$ is the vector of marginal proportions corresponding to A_j. Defining

$$\boldsymbol{\lambda}^{(j)}=P_j\boldsymbol{\pi}^{(j)} \qquad (j=1, \ldots, m)$$

by (3.17) and the standard rules for operations with Kronecker products, under H_0,

$$\begin{aligned}\boldsymbol{\lambda} &= (P_1 \times \cdots \times P_m)\boldsymbol{\pi} \\ &= (P_1 \times \cdots \times P_m)(\boldsymbol{\pi}^{(1)} \times \cdots \times \boldsymbol{\pi}^{(m)}) \\ &= P_1\boldsymbol{\pi}^{(1)} \times \cdots \times P_m\boldsymbol{\pi}^{(m)} = \boldsymbol{\lambda}^{(1)} \times \cdots \times \boldsymbol{\lambda}^{(m)} \end{aligned} \tag{3.21}$$

Since $\boldsymbol{\pi}^{(j)} = P_j^{-1}\boldsymbol{\lambda}^{(j)}$, retracing the steps above it may similarly be seen that (3.21) also implies H_0. Thus H_0 is equivalent to

$$H_0'\colon \boldsymbol{\lambda} = \boldsymbol{\lambda}^{(1)} \times \cdots \times \boldsymbol{\lambda}^{(m)}$$

Now introduce m pseudocharacters $B_1, \ldots, B_m$ such that for $j = 1, \ldots, m$, B_j has t_j categories $B_{j1}, \ldots, B_{jt_j}$, and an individual belongs to the category combination $(B_{1u_1}, \ldots, B_{mu_m})$ if his or her response is $(u_1, \ldots, u_m)$, $u_j = 1, \ldots, t_j$; $j = 1, \ldots, m$. Since H_0 and H_0' are equivalent, the hypothesis of complete independence of $A_1, \ldots, A_m$ is equivalent to the hypothesis of complete independence of $B_1, \ldots, B_m$. But $B_1, \ldots, B_m$ are observable and hence one can test H_0 by testing equivalently for complete independence of $B_1, \ldots, B_m$. In particular, for large samples, the standard Pearsonian χ^2 test may be applied as follows.

Let $\hat{\lambda}_{u_1 \cdots u_m}$ be a typical element of $\hat{\boldsymbol{\lambda}}$. Also, for $j = 1, \ldots, m$, let

$$\hat{\boldsymbol{\lambda}}^{(j)} = (\hat{\lambda}_1^{(j)}, \ldots, \hat{\lambda}_{t_j}^{(j)})^1$$

be the vector of sample proportions corresponding to $\boldsymbol{\lambda}^{(j)}$. Then the Pearsonian χ^2 test will be based on the statistic

$$\chi^2 = n \sum_{u_1=1}^{t_1} \cdots \sum_{u_m=1}^{t_m} \frac{\left(\hat{\lambda}_{u_1 \cdots u_m} - \prod_{j=1}^{m} \hat{\lambda}_{u_j}^{(j)}\right)^2}{\prod_{j=1}^{m} \hat{\lambda}_{u_j}^{(j)}} \tag{3.22}$$

and one should reject the null hypothesis of complete independence at level of significance α if the value of χ^2 exceeds $\chi^2_{\alpha,\nu}$, the upper α point of a chi-square variate with ν degrees of freedom, where $\nu = \prod_{j=1}^{m}(t_j - 1)$. In practice, the following computational formula is often helpful in a numerical evaluation of (3.22):

$$\chi^2 = n \sum_{u_1=1}^{t_1} \cdots \sum_{u_m=1}^{t_m} \frac{\hat{\lambda}^2_{u_1 \cdots u_m}}{\prod_{j=1}^{m} \hat{\lambda}_{u_j}^{(j)}} - n$$

In the development above, all the m attributes were taken into account. In particular situations, one may as well deal with subsets of attributes and test their independence in a similar fashion.

Generalizing the ideas of Eriksson (1973) and Bourke and Dalenius (1976), we present below a scheme for actually obtaining data using RR in multiattribute situations. In an SRSWR of size n, each respondent is asked about his A_j class and reports a true answer with probability p_j or reports $1, \ldots, t_j$ with respective probabilities $p_{j_1}, \ldots, p_{jt_j}, p_j + p_{j_1} + \cdots + p_{jt_j} = 1$; $j = 1, \ldots, m$, the reporting for different characters being done independently. Then, as in the Bourke and Dalenius (1976) scheme described in Section 3.3, the design matrices are given by

$$P_j = \begin{bmatrix} p_j + p_{j1} & p_{j1} & \cdots & p_{j1} \\ p_{j2} & p_j + p_{j2} & \cdots & p_{j2} \\ & \vdots & & \\ p_{jt_j} & p_{jt_j} & \cdots & p_j + p_{jt_j} \end{bmatrix} \qquad (j = 1, \ldots, m)$$

The rest of the analysis proceeds as usual.

Example 3.3. In a study concerning two attributes, let

A_1: alcoholism and A_2: practice of gambling with classes

A_{11}: consuming over 1000 ml

A_{12}: consuming between 250 and 1000 ml

A_{13}: consuming below 250 ml of alcoholic drinks over the last 1-month period, and

A_{21}: practicing any sort of gambling

A_{22}: not practicing gambling

over the last 1-month period. The RR scheme described in the preceding paragraph was adopted, the values of the design parameters being $p_1 = 0.25$, $p_{11} = p_{12} = p_{13} = 0.25$, $p_2 = 0.40$, and $p_{21} = p_{22} = 0.30$. Then $m = 2$, $t_1 = 3$, and $t_2 = 2$ and the design matrices are given by

$$P_1 = \begin{bmatrix} 0.50 & 0.25 & 0.25 \\ 0.25 & 0.50 & 0.25 \\ 0.25 & 0.25 & 0.50 \end{bmatrix} \qquad P_2 = \begin{bmatrix} 0.70 & 0.30 \\ 0.30 & 0.70 \end{bmatrix}$$

In an SRSWR of size $n = 100$, the (randomized) responses (1,1), (1,2), (2,1), (2,2), (3,1), and (3,2) were seen to have frequencies 13, 15, 16, 16, 16, and 24, respectively. Hence

$$\hat{\lambda} = (0.13,\ 0.15,\ 0.16,\ 0.16,\ 0.16,\ 0.24)'.$$

Now, as

$$P_1^{-1}\times P_2^{-1}=\begin{bmatrix}3.00 & -1.00 & -1.00\\ -1.00 & 3.00 & -1.00\\ -1.00 & -1.00 & 3.00\end{bmatrix}\times\begin{bmatrix}1.75 & -0.75\\ -0.75 & 1.75\end{bmatrix}$$

$$=\begin{bmatrix}5.25 & -2.25 & -1.75 & 0.75 & -1.75 & 0.75\\ -2.25 & 5.25 & 0.75 & -1.75 & 0.75 & -1.75\\ -1.75 & 0.75 & 5.25 & -2.25 & -1.75 & 0.75\\ 0.75 & -1.75 & -2.25 & 5.25 & 0.75 & -1.75\\ -1.75 & 0.75 & -1.75 & 0.75 & 5.25 & -2.25\\ 0.75 & -1.75 & 0.75 & -1.75 & -2.25 & 5.25\end{bmatrix}$$

by (3.18), $\boldsymbol{\pi}=(\pi_{11},\pi_{12},\pi_{21},\pi_{22},\pi_{31},\pi_{32})'$ is estimated unbiasedly by

$$\hat{\boldsymbol{\pi}}=(0.085,\ 0.035,\ 0.265,\ 0.015,\ 0.025,\ 0.575)'$$

Also,

$$\hat{\lambda}^{\delta}=\text{diag}(0.13,\ 0.15,\ 0.16,\ 0.16,\ 0.16,\ 0.24)'$$

and by (3.20), the dispersion matrix of $\hat{\boldsymbol{\pi}}$ is unbiasedly estimated by

$$\hat{\text{disp}}(\hat{\boldsymbol{\pi}})=(n-1)^{-1}[(P_1^{-1}\times P_2^{-1})\hat{\lambda}^{\delta}(P_1^{-1}\times P_2^{-1})'-\hat{\boldsymbol{\pi}}\,\hat{\boldsymbol{\pi}}']=$$

$$\begin{bmatrix}0.05596 & -0.04298 & -0.02611 & 0.01855 & -0.02772 & 0.02230\\ -0.04298 & 0.06259 & 0.01847 & -0.02538 & 0.02279 & -0.03549\\ -0.02611 & 0.01847 & 0.06320 & -0.04724 & -0.03171 & 0.02339\\ 0.01855 & -0.02538 & -0.04724 & 0.06644 & 0.02492 & -0.03729\\ -0.02772 & 0.02279 & -0.03171 & 0.02492 & 0.06754 & -0.05582\\ 0.02230 & -0.03549 & 0.02339 & -0.03729 & -0.05582 & 0.08291\end{bmatrix}$$

Turning to the problem of testing the null hypothesis of independence of the two attributes, observe that

$$\hat{\boldsymbol{\lambda}}^{(1)}=(0.13+0.15,\ 0.16+0.16,\ 0.16+0.24)'=(0.28,\ 0.32,\ 0.40)'$$

$$\hat{\boldsymbol{\lambda}}^{(2)}=(0.13+0.16+0.16,\ 0.15+0.16+0.24)'=(0.45,\ 0.55)'$$

and hence, by (3.22),

$$\chi^2=n\sum_{u_1=1}^{3}\sum_{u_2=1}^{2}\frac{(\hat{\lambda}_{u_1u_2}-\hat{\lambda}_{u_1}^{(1)}\hat{\lambda}_{u_2}^{(2)})^2}{\hat{\lambda}_{u_1}^{(1)}\hat{\lambda}_{u_2}^{(2)}}$$

$$=0.7504$$

Here $v=(t_1-1)(t_2-1)=2$, and taking the level of significance as 0.05, the computed value of χ^2 is less than $\chi^2_{0.05,2}$ $(=5.99)$. Hence the null hypothesis of independence appears to be acceptable at the 0.05 level of significance. It may, however, be noted that the test for independence in an RR setting is not very powerful, and one should, therefore, interpret the result with due caution.

Tamhane (1981) presents an alternative plan for m dichotomous attributes $A_1, \ldots, A_m$. Suppose that interest lies only in the marginal and bivariate proportions, and for $1 \le j_i < j_2 \le m$, let

$$\theta_{j_1} = \text{unknown proportion of individuals possessing } A_{j_1}$$

$$\theta_{j_1 j_2} = \text{unknown proportion of individuals possessing both } A_{j_1} \text{ and } A_{j_2}$$

The randomization device contains m statements of which the first s announce membership in $A_j (j=1, \ldots, s)$ and the remaining $m-s$ decline membership in $A_j (j=s+1, \ldots, m)$, an appropriate choice of s being $s \simeq m/2$. For $1 \le j_1 < j_2 \le m$, let ϕ_{j_1} and $\phi_{j_1 j_2}$ be defined in the same manner as θ_{j_1} and $\theta_{j_1 j_2}$ but with respect to the modified attributes B_j which are either the original $A_j (j=1, \ldots, s)$ or their complements $(j=s+1, \ldots, m)$. Since the θ's are one-to-one functions of the ϕ's, it will be enough to consider the problem in terms of the ϕ's.

An SRSWR of size n is split up into b subsamples of sizes $n_1, \ldots n_b (\Sigma_{i=1}^{b} n_i = n)$. Each respondent is presented with all the m statements; he or she picks up one statement at random (as before, unobserved by the interviewer) and reports accordingly. The procedure is repeated once again independently. Let $p_{ij}^{(h)}$ be the probability that an individual in the ith subsample encounters the jth statement on the hth trial $(i=1, \ldots, b; j=1, \ldots, m; h=1,2)$.

Suppose the responses are coded so that a score of 2^{h-1} is assigned to a "yes" response on the hth trial and a score of 0 is assigned to a "no" response. For each interviewee, there are then four possible total scores—0, 1, 2, and 3—as shown in Table 3.2. Then denoting by λ_{iv} the probability of obtaining a total score v for an interviewee in the ith subsample and writing

$$\boldsymbol{\lambda} = (\lambda_{11}, \lambda_{12}, \lambda_{13}, \ldots, \lambda_{b1}, \lambda_{b2}, \lambda_{b3})'$$
$$\boldsymbol{\phi} = (\phi_1, \ldots, \phi_m, \phi_{12}, \ldots, \phi_{m-1,m})'$$

it can be shown that

$$\boldsymbol{\lambda} = R\boldsymbol{\phi} \tag{3.23}$$

TABLE 3.2

Response in:		
First trial	Second trial	Total score
No	No	0+0=0
Yes	No	1+0=1
No	Yes	0+2=2
Yes	Yes	1+2=3

where the elements of the matrix $R=((R_{uj}))$ are as follows. For $i=1, \ldots, b$; $j=1, \ldots, m$,

$$R_{3i-2,j}=p_{ij}^{(1)}(1-p_{ij}^{(2)}) \qquad R_{3i-1,j}=p_{ij}^{(2)}(1-p_{ij}^{(1)}) \qquad R_{3i,j}=p_{ij}^{(1)}p_{ij}^{(2)} \tag{3.24}$$

and for $1 \leq j_1 < j_2 \leq m$, if $k=j_1m-j_1(j_1+1)/2+j_2$, then

$$R_{3i-2,k}=R_{3i-1,k}=-R_{3i,k}=-(p_{ij_1}^{(1)}p_{ij_2}^{(2)}+p_{ij_2}^{(1)}p_{ij_1}^{(2)}) \tag{3.25}$$

In Exercise 3.5 we ask the reader to prove (3.24) and (3.25).

Tamhane (1981) argues that by suitable choice of the $p_{ij}^{(h)}$'s, the matrix R can be made to have full column rank only if $b \geq \binom{m}{2}$. Assume that $b \geq \binom{m}{2}$ and that R has full column rank. Let n_{iv} be the number of individuals in the ith subsample having a total score v ($i=1, \ldots, b$; $v=0,1,2,3$). Then one can obtain the maximum likelihood estimate of $\boldsymbol{\phi}$ by directly maximizing the likelihood function

$$L=\text{constant} \times \prod_{i=1}^{b}\prod_{v=0}^{3}(\lambda_{iv})^{n_{iv}} \tag{3.26}$$

subject to the natural constraints

$$0 \leq \phi_j \leq 1, \qquad (j=1, \ldots, m)$$

and

$$\max(0, \phi_{j_1}+\phi_{j_2}-1) \leq \phi_{j_1 j_2} \leq \min(\phi_{j_1}, \phi_{j_2}) \qquad (1 \leq j_1 < j_2 \leq m) \tag{3.27}$$

In (3.26), the λ_{iv}'s (v = 1, 2, 3) are given in terms of $\boldsymbol{\phi}$ by (3.23), while

$$\lambda_{i0}=1-\lambda_{i1}-\lambda_{i2}-\lambda_{i3} \qquad (i=1, \ldots, b)$$

The constraints in (3.27) are linear and the objective function log L may be seen to be concave in the elements of $\boldsymbol{\phi}$. The resulting nonlinear programming problem is, therefore, well structured and, as Tamhane (1981) indicates, can be solved quite economically on a computer. Tamhane also treats some aspects of the choice of design parameters, keeping in view the question of respondent privacy.

EXERCISES

3.1. Considering the second approach of Eriksson (1973) described in Section 3.2, derive a UE of $\boldsymbol{\pi}=(\pi_1, \ldots, \pi_s)'$. Observe that this is possible without any explicit matrix inversion. Obtain the dispersion matrix of the estimator and suggest a UE of this dispersion matrix.

3.2. Perform the algebraic steps to show that the estimator in (3.15) is the same as that obtained through direct inversion of the design matrix P and use of (3.10).

3.3. (Bourke, 1974) For a trichotomous sensitive attribute with unknown proportions, π_1, π_2, and π_3, the following RR scheme, involving two stages of randomization, is proposed. Together with the sensitive character, consider a trichotomous unrelated character U, the true proportions μ_1, μ_2, and μ_3 corresponding to U being known. An SRSWR of size n is drawn. In the first stage of randomization, each interviewee draws a card at random from a deck made up of three card types as shown in Figure 3.3. As usual, S_j and

	card type 1	card type 2	card type 3
If bead is blue			
Answer (1)	S_1	U_1	S_3
from (2)	S_2	U_2	S_1
here (3)	S_3	U_3	S_2
If bead is red			
Answer (1)	U_1	S_2	U_1
from (2)	U_2	S_3	U_2
here (3)	U_3	S_1	U_3

FIGURE 3.3

U_j are abbreviations for the statements "I belong to sensitive category j" and "I belong to unrelated category j," respectively. In the second stage, a box of beads coloured blue and red in proportions q_b and $q_r = 1 - q_b$ is used to choose from which portion of the card the respondent should reply. Let a_1, a_2, and a_3 be the proportions of the three types of cards in the deck. Show that the probabilities λ_1, λ_2, and λ_3 of responses 1, 2, and 3 are given by

$$\lambda_1 = a_1(q_b\pi_1 + q_r\mu_1) + a_2(q_b\mu_1 + q_r\pi_2) + a_3(q_b\pi_3 + q_r\mu_1)$$
$$\lambda_2 = a_1(q_b\pi_2 + q_r\mu_2) + a_2(q_b\mu_2 + q_r\pi_3) + a_3(q_b\pi_1 + q_r\mu_2)$$
$$\lambda_3 = a_1(q_b\pi_3 + q_r\mu_3) + a_2(q_b\mu_3 + q_r\pi_1) + a_3(q_b\pi_2 + q_r\mu_3)$$

Hence suggest a UE for $\pi = (\pi_1, \pi_2, \pi_3)'$.

3.4. (Bourke, 1982) Consider a situation with two attributes each with two classes. Let π_{ij} be the proportion in the (i, j)th cell $(i, j = 1, 2)$. In an SRSWR of size n, each respondent is asked to choose with specified probabilities P_i $(i = 1, 2, 3, 4; \Sigma P_i = 1)$ one of the four alternatives, as in Table 3.3, and report accordingly. Suggest a UE for $\pi = (\pi_{11}, \pi_{12}, \pi_{21}, \pi_{22})'$. Obtain the dispersion matrix of your estimator and a UE for the dispersion matrix.

3.5. With reference to (3.19), show that $\mathrm{disp}(\hat{\pi})$ can be expressed as $\mathrm{disp}(\hat{\pi}) = \Sigma_1 + \Sigma_2$, where

$$\Sigma_1 = n^{-1}(\pi^\delta - \pi\pi') \qquad \Sigma_2 = n^{-1}P^{-1}(\lambda^\delta - P\pi^\delta P')P'^{-1}$$

Verify that Σ_1 and Σ_2 are both nonnegative definite and interpret this decomposition. *Hint*: To prove the nonnegative definiteness of Σ_2, one may employ Lemma 3.1 provided that P has each column sum

TABLE 3.3

True class	Report under alternative: 1	2	3	4
11	1	4	3	2
12	2	1	4	3
21	3	2	1	4
22	4	3	2	1

unity. To check this condition, let $\underline{1}_j$ be a t_j-component vector with each element unity and $\underline{1} = \underline{1}_1 \times \cdots \times \underline{1}_m$. Since $P = P_1 \times \cdots \times P_m$, by the standard rules for operation with Kronecker products, $P'\underline{1} = (P'_1\underline{1}_1) \times \cdots \times (P'_m\underline{1}_m)$. Then apply (3.16).

3.6. Verify the relations (3.24) and (3.25).

REFERENCES

Abul-Ela, Abdel-Latif, A., Greenberg, B. G., and Horvitz, D. G. (1967). A multiproportions RR model. *J. Amer. Statist. Assoc.* **62**, 990–1008.

Bourke, P. D. (1974). Multiproportions RR using the unrelated question technique. Private communication.

Bourke, P. D. (1978). RR designs with symmetric response for multiproportions situations. *Statist. Tidskrift* **16**, 197–207.

Bourke, P. D. (1981). On the analysis of some multivariate RR designs for categorical data. *J. Statist. Plann. Inference* **5**, 165–170.

Bourke, P. D. (1982). RR multivariate designs for categorical data. *Comm. Statist.— Theory Methods* **11**, 2889–2901.

Bourke, P. D., and Dalenius, T. (1976). Some new ideas in the realm of randomized inquiries. *Internat. Statist. Rev.* **44**, 219–221.

Drane, W. (1976). Theory of RR to two sensitive questions. *Comm. Statist— Theory Methods* **5**, 565–575.

Eriksson, S. (1973). Randomized interviews for sensitive questions. Ph.D. thesis, University of Gothenburg.

Greenberg, B. G., Abul-Ela, Abdel-Latif, A., Simmons, W. R., and Horvitz, B. G. (1969). The unrelated question RR model: theoretical framework. *J. Amer. Statist. Assoc.* **64**, 520–539.

Liu, P. T., and Chow, L. P. (1976). A new discrete quantitative RR model. *J. Amer. Statist. Assoc.* **71**, 72–73.

Mukerjee, R. (1981). Inference on confidential characters from survey data. *Bull. Calcutta Statist. Assoc.* **30**, 77–88.

Mukhopadhyay, P. (1980). On estimation of some confidential characters from survey data. *Bull. Calcutta Statist. Assoc.* **29**, 133–141.

Tamhane, A. C. (1981). RR techniques for multiple sensitive attributes. *J. Amer. Statist. Assoc.* **76**, 916–923.

Appendix

Supplementary Remarks on the Polychotomous and Multiattribute Models

A3.1 AUGMENTATION MODELING

Let us first consider an additive model presented by Kim and Flueck (1978). Let there be t mutually exclusive and exhaustive categories, with respective proportions $\pi_1, \pi_2, \ldots, \pi_t$, according to a sensitive character A. For each person in the population, let X be a variable assuming the value j if the person belongs to the jth category. In an SRSWR of size n, each respondent is asked to choose, independently of his or her X value, a number i ($i=1, \ldots, t$) with an assigned probability P_i ($\Sigma P_i=1$). Let Y denote the number chosen and define

$$\begin{aligned} R &= X+Y && \text{if} \quad X+Y \leq t \\ &= X+Y-t && \text{if} \quad X+Y > t \end{aligned}$$

Each respondent is asked to report the value R.

The probability distribution of R, in terms of $\pi_1, \pi_2, \ldots, \pi_t$, is easy to determine. For example, if $t=3$, then

$$\begin{aligned} \lambda_1 = P(R=1) &= P(X=1,\ Y=3)+P(X=2,\ Y=2)+P(X=3,\ Y=1) \\ &= \pi_1 P_3 + \pi_2 P_2 + \pi_3 P_1 \end{aligned}$$

and similarly, $\lambda_2 = P(R=2)$ and $\lambda_3 = P(R=3)$ may be written down.

Then with

$$\boldsymbol{\lambda}^* = (\lambda_1 - P_1, \lambda_2 - P_2)', \qquad \boldsymbol{\pi} = (\pi_1, \pi_2)'$$

$$P = \begin{pmatrix} P_3 - P_1 & P_2 - P_1 \\ P_1 - P_2 & P_3 - P_2 \end{pmatrix}$$

one may note that

$$\boldsymbol{\lambda}^* = P\boldsymbol{\pi}$$

Let P_1, P_2, and P_3 be not all equal to make P nonsingular. Then one gets a UE for $\boldsymbol{\pi}$ as $\hat{\boldsymbol{\pi}} = P^{-1}\hat{\boldsymbol{\pi}}^*$, where

$$\hat{\boldsymbol{\lambda}}^* = (\hat{\lambda}_1 - P_1, \hat{\lambda}_2 - P_2)'$$

and $\hat{\lambda}_i$ is the sample proportion corresponding to λ_i, $i = 1, 2$. Also, π_3 is estimated by $\hat{\pi}_3 = 1 - (\hat{\pi}_1 + \hat{\pi}_2)$. The dispersion matrix for $\hat{\boldsymbol{\pi}}$ and a UE thereof may be obtained following the line of Section 3.3. The case $k > 3$ can be similarly covered.

A3.2 TWO-STAGE SCHEMES

The need for reckoning with the importance of distinguishing the stigmatizing groups from the innocent ones in multiproportion problems was emphasized by Hochberg (1975). He considered estimating t group proportions of which $r(< t - 1)$ groups may be supposed to be stigmatizing. He recommended a two-stage randomized response scheme. Suppose that the first r groups are stigmatizing. In the first stage a sampled person is asked with a probablility P_i to say truthfully whether or not he or she belongs to a sensitive group i, and with probability $1 - P_i$ the person is asked to say "yes"; thus this is a contamination design. The procedure is repeated r times independently for each sample person, to cover r sensitive groups.

Suppose that out of an SRSWR of n individuals a "no" response comes from n^0 persons in respect of each of the r questions. Further, let $\pi_1, \ldots, \pi_t$ be the true proportions in the t groups. Then $\lambda_i = P_i\pi_i + (1 - P_i)$ is the probability of a "yes" response about i $(i = 1, \ldots, r)$. Writing $\hat{\lambda}_i$ as the corresponding sample proportion, a UE for π_i is

$$\hat{\pi}_i = P_i^{-1}[\hat{\lambda}_i - (1 - P_i)] \qquad (i = 1, \ldots, r)$$

Since the n^0 individuals bear no stigma, at the second stage each may be asked directly to reveal the group out of $(r + 1), \ldots, t$ to which he or she belongs. Let n_j^0 be the sampled number of persons (out of n^0 persons) who declare that they belong to the jth group and write

$$\hat{\pi}_j^* = \frac{n_j^0}{n^0}$$

which yields a UE for π_j as

$$\hat{\pi}_j = \hat{\pi}_j^* \left(1 - \sum_{j=1}^{r} \hat{\pi}_j\right) \qquad (j = r+1, \ldots, t)$$

Hochberg (1975) noted the need for an asymptotic analysis to derive the formula for variances of $\hat{\pi}_j \, (j = r+1, \ldots, t)$. He also presents some other schemes for polychotomous situations.

A3.3 SOME REMARKS

Bourke (1982) uses Simmons' unrelated-question models and applies the schemes of Bourke and Dalenius (1976) in conjunction with Bourke's symmetry-of-response concept in deriving UEs for multiattribute proportions. For the sake of simplicity he treats only the cases of known parameters concerning the unrelated questions. Finally, we mention that Chen (1978) considers an application of log-linear models for multiattribute data obtained through RR.

REFERENCES

Bourke, P. D. (1974). Multi-proportions RR using the unrelated question technique. Private communication.

Bourke, P. D. (1982). RR multivariate designs for categorical data. *Comm. Statist.—Theory Methods* **11**, 2889–2901.

Bourke, P. D., and Dalenius, T. (1976). Some new ideas in the realm of randomized inquiries. *Internat. Statist. Rev.* **44**, 219–221.

Chen, T. T. (1978). Log-linear models for the categorical data obtained from randomized response techniques. *Proc. ASA Soc. Statist. Sec.*, 284–288.

Hochberg, Y. (1975). Two stage randomized response schemes for estimating a multinomial. *Comm. Statist.—Theory Methods* **4**, 1021–1032.

Kim, Jong-Ik, and Flueck, John A. (1978). An additive randomized response model. *Proc. ASA Sec. Surv. Res. Methods*, 351–355.

4

Techniques for Quantitative Characters

4.1. INTRODUCTION

In the preceding chapters we described some RR methods for attributes or sensitive characters which are essentially qualitative in nature. In such situations, there is a natural partitioning of the population into mutually exclusive and exhaustive classes according to the forms of the attribute(s) under consideration, and interest lies in the estimation of the respective proportions in these classes. In practice, however, one may also have to deal with quantitative sensitive characters or variables. For example, one may be interested in investigating the numbers of induced abortions among the females in a region or the quantity of alcoholic drinks consumed by a given group of people over a given period.

In this chapter we first examine how the unrelated-question model described earlier can be suitably modified to deal with variables. This is followed by a description of some other procedures for quantitative characters. Finally, recognizing the problem of estimation in RR models as one in mixture models, it will be seen how the available theories of Hoeffding's U statistic or Von Mises' differentiable statistical functions can be modified to derive optimal estimators in the RR setup.

4.2. THE UNRELATED-QUESTION MODEL

Greenberg *et al.* (1971) suggest an extension of the unrelated-question model to quantitative characters. Consider a sensitive variable X which is supposed to be continuous with true density $g(\cdot)$ (the treatment remains similar for a discrete X if one replaces everywhere a density function by the corresponding probability mass function). Let Y be an unrelated character with density $h(\cdot)$. The problem is to estimate μ_X, the unknown population mean of X.

Considering first the simple case when μ_Y, the population mean of Y, is known, suppose that an SRSWR is drawn and each interviewee reports his or her X value with probability P and Y value with $Q = 1 - P$. Then an RR, say Z, has density

$$f(z) = Pg(z) + Qh(z) \tag{4.1}$$

Hence

$$E(Z) = PE(X) + QE(Y) \qquad E(Z^2) = PE(X^2) + QE(Y^2)$$

and consequently, the population mean, μ_Z, and the population variance, σ_Z^2, of Z are given by

$$\mu_Z = P\mu_X + Q\mu_Y \tag{4.2}$$

$$\sigma_Z^2 = E(Z^2) - [E(Z)]^2 = P\sigma_X^2 + Q\sigma_Y^2 + PQ(\mu_X - \mu_Y)^2 \tag{4.3}$$

where σ_X^2 and σ_Y^2 are the population variances of X and Y, respectively.

From an SRSWR of size n, let the available (randomized) responses be $Z_1, \ldots, Z_n$. Defining the sample mean and the sample variance

$$\bar{Z} = n^{-1}\textstyle\sum_{i=1}^{n} Z_i \qquad s_Z^2 = (n-1)^{-1}\textstyle\sum_{i=1}^{n}(Z_i - \bar{Z})^2$$

one has

$$E(\bar{Z}) = \mu_Z \tag{4.4a}$$

$$E(s_Z^2) = \sigma_Z^2 \tag{4.4b}$$

By (4.2) and (4.4a), one obtains a UE of μ_X as

$$\hat{\mu}_X = \frac{\bar{Z} - Q\mu_Y}{P}$$

with

$$\text{var}(\hat{\mu}_X) = \frac{\text{var}(\bar{Z})}{P^2} = \frac{\sigma_Z^2}{nP^2}$$

By (4.4b), a UE of $\text{var}(\hat{\mu}_X)$ is given by $\widehat{\text{var}}(\hat{\mu}_X) = s_Z^2/nP^2$.

In the more realistic situation when μ_Y is unknown, two independent SRSWRs of sizes n_1 and n_2 are drawn. For $i = 1,2$, a respondent in the ith sample reports his or her X value with probability P_i and Y value with $Q_i = 1 - P_i$, and the RR Z has the density

$$f_i(z) = P_i g(z) + Q_i h(z) \tag{4.5}$$

Hence if $\bar{Z}_1$ and $\bar{Z}_2$ are the sample means of the Z values, it follows as before that

$$E(\bar{Z}_i) = P_i\mu_X + Q_i\mu_Y$$

whence provided that $P_1 \neq P_2$, a UE of μ_X is obtained as

$$\hat{\mu}_X = \frac{(1-P_2)\bar{Z}_1 - (1-P_1)\bar{Z}_2}{P_1 - P_2} \qquad (4.6)$$

which is simply a modified version of (2.5) for quantitative characters.

Since $\bar{Z}_1$ and $\bar{Z}_2$ are independent, it is straightforward to check that

$$\mathrm{var}(\hat{\mu}_X) = \frac{[(1-P_2)^2\sigma_1^2/n_1] + [(1-P_1)^2\sigma_2^2/n_2]}{(P_1-P_2)^2}$$

where, analogously to (4.3),

$$\sigma_i^2 = P_i\sigma_X^2 + Q_i\sigma_Y^2 + P_iQ_i(\mu_X - \mu_Y)^2 \qquad (i=1,2)$$

is the variance of a random variable having density (4.5). A UE of $\mathrm{var}(\hat{\mu}_X)$ is given by

$$\widehat{\mathrm{var}}(\hat{\mu}_X) = \frac{[(1-P_2)^2 s_1^2/n_1] + [(1-P_1)^2 s_2^2/n_2]}{(P_1-P_2)^2} \qquad (4.7)$$

where s_i^2 is the variance, with divisor $n_i - 1$, of the available observations in the ith sample ($i=1,2$).

As in Section 2.4, an appropriate choice of the design parameters n_1, n_2, P_1 and P_2 is needed. If the total sample size n $(=n_1+n_2)$ is fixed, then, analogously to (2.8), $\mathrm{var}(\hat{\mu}_X)$ is minimum when

$$\frac{n_1}{n_2} = \frac{Q_2\sigma_1}{Q_1\sigma_2}$$

Also, as in the case of attributes, for minimizing $\mathrm{var}(\hat{\mu}_X)$ one may take $P_2=0$ and P_1 as large as possible without prejudice to maintenance of privacy. Further, for fixed (P_1, P_2, n_1, n_2), it may be seen that $\mathrm{var}(\hat{\mu}_X)$ is decreasing in $|\mu_X - \mu_Y|$ and also in σ_Y^2. Therefore, the unrelated character Y should be chosen such that μ_Y is close to μ_X and σ_Y^2 is small. However, too small a choice of σ_Y^2, compared to σ_X^2, will meet with suspicion and discourage faithful cooperation. Therefore, as a working rule, keeping both efficiency and protection of privacy in mind, a choice of σ_Y^2 as close as possible to σ_X^2 from above, without dissuading responses, should be attempted.

If θ is a parameter concerning the distribution of X, it is easy to see that for a UE $\hat{\theta}$ of θ, based on the randomized responses,

$$\operatorname{var}(\hat{\theta}) \geq \left\{ n_1 E_\theta \left[\frac{d}{d\theta} \log f_1(Z) \right]^2 + n_2 E_\theta \left[\frac{d}{d\theta} \log f_2(Z) \right]^2 \right\}^{-1} \tag{4.8}$$

which is a form of the Rao-Cramér inequality, adapted to the present situation. The efficiency of an arbitrary UE $\hat{\theta}^*$ of θ may be defined as the ratio of the bound (4.8) to $\operatorname{var}(\hat{\theta}^*)$. Considering particular forms of $g(\cdot)$ and $h(\cdot)$ as Poisson, exponential, and normal, and taking $n_1 = n_2$, Greenberg *et al.* (1971) exhibit the efficiency of the estimator $\hat{\mu}_X$ specified in (4.6). In Exercise 4.2 we ask the reader to work out some further details in this connection.

Example 4.1. In a survey to estimate the average monthly expenditure on alcoholic drinks of undergraduate college students in Calcutta, two independent SRSWRs of sizes 20 and 25 were taken. Each interviewee was required to answer one of the following two questions, chosen on a probability basis:

1. How much did you spend on alcoholic drinks over the last 1-month period?
2. How much did you spend on purchasing books related to your subjects of study over the last 1-month period?

For randomization, two decks of cards, one for each sample, were used. The deck meant for the first sample contained 50 cards, of which 30 and 20 bore questions 1 and 2, respectively. The corresponding figures for the deck meant for the second sample were 60, 24, and 36. As usual, each respondent was requested to draw one card from the deck supplied and then reply to the question on the card drawn. The (randomized) responses were as follows:

Sample 1: 7, 15, 18, 10, 20, 30, 0, 10, 8, 6, 15, 12, 0, 5, 20, 12, 10, 25, 18, 16
Sample 2: 15, 20, 8, 7, 25, 0, 22, 18, 16, 15, 5, 12, 10, 16, 10, 12, 20, 5, 24, 20, 18, 6, 20, 15, 16

all the figures being in the Indian rupee.

With the data above, $\bar{Z}_1 = 12.85$, $\bar{Z}_2 = 14.2$, $s_1^2 = 59.9237$, and $s_2^2 = 42.5833$. Also, $n_1 = 20$, $n_2 = 25$, $P_1 = 0.6$, and $P_2 = 0.4$. Hence by (4.6) and (4.7), an unbiased estimate of μ_X, the mean monthly expenditure on alcoholic drinks, is $\hat{\mu}_X = 10.15$ with $\widehat{\operatorname{var}}(\hat{\mu}_X) = 33.7790$.

4.3. SOME ADDITIONAL TECHNIQUES

Dalenius and Vitale (1974) describe an RR procedure for sensitive quantitative characters. First, suppose that the study variable X is discrete, assuming values $1, 2, \ldots, t$ with respective unknown relative frequencies $\pi_1, \pi_2, \ldots, \pi_t$ in the population. To estimate the mean, say $\mu = \Sigma_{j=1}^{t} j\pi_j$, of X, an SRSWR of size n is drawn. Each respondent is provided with a spinner fitted with a printer. By drawing angles at the center, the face of the spinner is divided into t equal sectors labeled $0, 1, \ldots, t-1$ (Figure 4.1 shows the face of the spinner for $t=6$). Each respondent is asked to spin the spinner and report whether or not the number indicated by the pointer is less than his or her X value. Then the probability of a "yes" response is

$$\lambda = \sum_{j=1}^{t} \frac{j}{t} \pi_j = \mu t^{-1}$$

Hence if $\hat{\lambda}$ is the sample proportion of "yes" responses, a UE of μ is given by $\hat{\mu} = t\hat{\lambda}$, with $\text{var}(\hat{\mu}) = t^2\lambda(1-\lambda)/n$. A UE of $\text{var}(\hat{\mu})$ is provided obviously by $t^2\hat{\lambda}(1-\hat{\lambda})/(n-1)$.

Now considering the continuous case, let X assume values in an interval $[0, t]$ with density $g(x)$ (the more general interval $[a, a+t]$ may be treated similarly). The procedure mimics that in the discrete case, this time

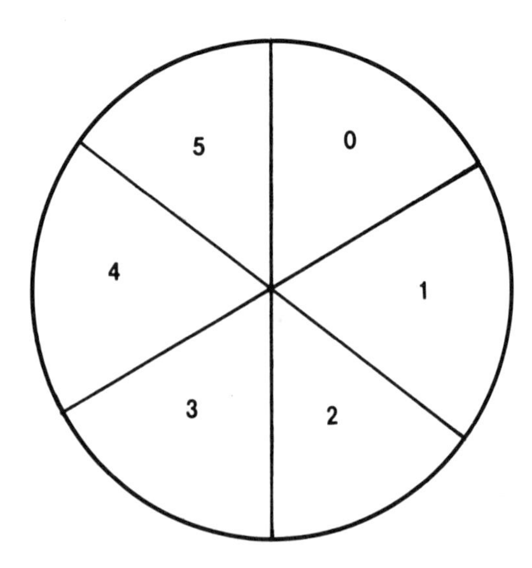

FIGURE 4.1

the face of the spinner being labeled continuously from 0 to t. Defining λ as before, it follows that

$$\lambda = \int_0^t \frac{x}{t} g(x)dx = \mu t^{-1}$$

and the rest of the analysis proceeds as before.

Liu *et al.* (1975) describe another technique for estimating the mean of a quantitative character. The method is essentially one suited for the multiproportions case. Let X be a sensitive discrete variable assuming values $x_1, \ldots, x_t$ with respective unknown true proportions $\pi_1, \ldots, \pi_t$. The randomization device looks like a flask with a body and a long neck, as shown in Figure 4.2. The device contains balls of t different colors such that the balls of the jth color represent the individuals with variate values x_j $(j = 1, \ldots, t)$.

Suppose that there are m_j balls of the jth color and let $m = \Sigma_{j=1}^{t} m_j$. The respondent is asked to turn the device upside down, shake it thoroughly, and allow all the balls to fall into the neck. The neck has m locations serially

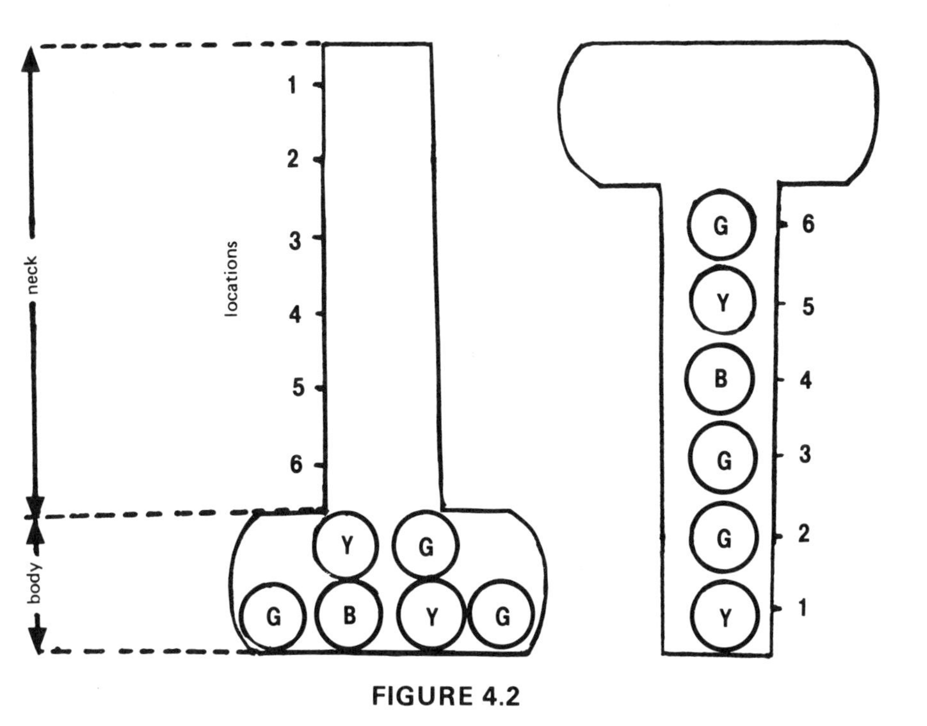

FIGURE 4.2

marked from the bottom. The respondent reports only the "bottom-most" location occupied by a ball of the color representing his or her variate value. Then the probability of the response i is

$$\lambda_i = \sum_{j=1}^{t} p_{ij}\pi_j \qquad (i=1, \ldots, w) \tag{4.9}$$

where $w = \max_i (m - m_i + 1)$ and p_{ij} is the probability that an individual with variate value x_j reports i. It is easy to see that (see Exercise 4.3)

$$p_{ij} = \frac{(m-m_j)!\ (m-i)!\ m_j}{m!\ (m-m_j-i+1)!} \tag{4.10}$$

With an SRSWR of size n let $\hat{\lambda}_i$ be the sample relative frequency of the response i. Since $w > t$, a UE of $\pi = (\pi_1, \ldots, \pi_t)'$, based on a substitution of the $\hat{\lambda}_i$'s in place of the λ_i's in (4.9), will not be unique. Liu *et al.* (1975), therefore describe a method of least squares in this connection. Since $\Sigma_{j=1}^{t}\pi_j = 1$, it follows from (4.9) that

$$E(\hat{\lambda}_i) = \lambda_i = \sum_{j=1}^{t-1}(p_{ij} - p_{it})\pi_j + p_{it} \qquad (i=1, \ldots, w) \tag{4.11}$$

Also, the dispersion matrix of $\hat{\lambda} = (\hat{\lambda}_1, \ldots, \hat{\lambda}_w)$ is $V = ((v_{ii'}))$, where $v_{ii} = \lambda_i(1-\lambda_i)/n$, $v_{ii'} = -\lambda_i\lambda_{i'}/n$ $(i \neq i')$. Hence defining

$$\underset{\sim}{Y} = (\hat{\lambda}_1 - p_{1t}, \ldots, \hat{\lambda}_w - p_{wt})' \tag{4.12}$$

one may take the linear model

$$\underset{\sim}{Y} = X\pi^* + \underset{\sim}{e}$$

where

$$\pi^* = (\pi_1, \ldots, \pi_{t-1})'$$

$$X = \begin{bmatrix} p_{11} - p_{1t} & p_{12} - p_{1t} & \cdots & p_{1t-1} - p_{1t} \\ & \vdots & & \\ p_{w1} - p_{wt} & p_{w2} - p_{wt} & \cdots & p_{wt-1} - p_{wt} \end{bmatrix} \tag{4.13}$$

$$\underset{\sim}{e} = (e_1, \ldots, e_w)'$$

with $E(\underset{\sim}{e}) = 0$ and disp$(\underset{\sim}{e}) = V$. Now V is positive semidefinite and the column space of X is a subspace of the column space of V. Hence by the method of weighted least squares (see, e.g., Rao, 1973, p. 300) one may consider the equations

$$X'V^- X\pi^* = X'V^- \underset{\sim}{Y} \tag{4.14}$$

where V^- is any generalized inverse of V. In particular, a simple choice of V^- is given by

$$V^- = n\,\text{diag}(\lambda_1^{-1}, \ldots, \lambda_w^{-1}) \tag{4.15}$$

Assume that $m_1, \ldots, m_t$ are such that X has full column rank. Then $X'V^-X$ is nonsingular, and had V been known, (4.14) would have given the best linear unbiased estimator of $\boldsymbol{\pi}^*$. In the present situation, however, V involves the λ_i's which are unknown, and therefore, one has to solve (4.14) for $\boldsymbol{\pi}^*$ by an iterative method as described below.

Let $\boldsymbol{\lambda} = (\lambda_1, \ldots, \lambda_w)'$, $G = G(\boldsymbol{\lambda}) = \text{diag}(\lambda_1^{-1}, \ldots, \lambda_w^{-1})$, and $\boldsymbol{\xi} = (p_{1t}, \ldots, p_{wt})'$. Then (4.11) may be expressed in matrix notation as

$$\boldsymbol{\lambda} = X\boldsymbol{\pi}^* + \boldsymbol{\xi} \tag{4.16}$$

Also, with V^- as in (4.15), equation (4.14) becomes

$$X'GX\boldsymbol{\pi}^* = X'G\underline{Y} \tag{4.17}$$

Based on (4.16) and (4.17), the iterative steps will be as follows. Set $\boldsymbol{\lambda}^{(0)} = (\hat{\lambda}_1, \ldots, \hat{\lambda}_w)'$, and compute

$$\hat{\boldsymbol{\pi}}^{*(u)} = [X'G(\boldsymbol{\lambda}^{(u)})X]^{-1}X'G(\boldsymbol{\lambda}^{(u)})\underline{Y}$$
$$\boldsymbol{\lambda}^{(u+1)} = X\hat{\boldsymbol{\pi}}^{*(u)} + \boldsymbol{\xi} \qquad (u = 0,1,2, \ldots)$$

The procedure continues until the elements of $\hat{\pi}^{*(u)}$ stabilize up to the required number of decimal places. The solution gives an estimator $\hat{\boldsymbol{\pi}}^* = (\hat{\pi}_1, \ldots, \hat{\pi}_{t-1})'$, say, of $\boldsymbol{\pi}^*$, from which π_t may be estimated as $\hat{\pi}_t = 1 - \Sigma_{j=1}^{t-1}\hat{\pi}_j$. Finally, $\mu = \Sigma_{j=1}^{t} x_j\pi_j$, the population mean, is estimated as

$$\hat{\mu} = \sum\nolimits_{j=1}^{t} x_j\hat{\pi}_j$$

Example 4.2. This example follows the line of the one described in Liu *et al.* (1975). In a survey to study the pre- and extramarital sexual behavior of the people in a region with individuals of opposite sexes other than legally married spouses, the sensitive question was phrased as follows: "With how many partners other than your legally married spouse have you had sexual relations in your lifetime?"

The randomization device, as in Figure 4.2, contained 6 balls (1 black, 2 yellow, and 3 green). The black, yellow, and green colors represented the following alternatives:

1. "I have never had such an experience."
2. "I have had such an experience with one partner."
3. "I have had such an experience with two or more partners."

Then $t=3$, $m_1=1$, $m_2=2$, $m_3=3$, $m=6$, $w=6$ and the values of the p_{ij}'s, obtained from (4.10), are as shown in Table 4.1. An SRSWR of size 300 was taken and the observed relative frequencies of the (randomized) responses were as shown in Table 4.2.

By (4.12) and (4.13),

$$\underline{Y}=\tfrac{1}{300}\,(-68,\ -22,\ 8,\ 25,\ 33,\ 24)'$$

$$X=\tfrac{1}{60}\begin{bmatrix} -20 & -10 \\ -8 & -2 \\ 1 & 3 \\ 7 & 5 \\ 10 & 4 \\ 10 & 0 \end{bmatrix}$$

Also, $\xi=(1/20)\,(10, 6, 3, 1, 0, 0)'$. The iterative steps for solving (4.14) are as shown below:

$$\lambda^{(0)}=(\hat{\lambda}_1, \ldots, \hat{\lambda}_6)'=\tfrac{1}{300}\,(82, 68, 53, 40, 33, 24)'$$

$$\hat{\pi}^{*(0)}=[X'G(\lambda^{(0)})X]^{-1}\,X'G(\lambda^{(0)})\underline{Y}=(0.486,\ 0.377)'$$

$$\lambda^{(1)}=X\hat{\pi}^{*(0)}+\xi$$

$$=(0.27533, 0.22270, 0.17694, 0.13806, 0.10605, 0.08092)'$$

$$\hat{\pi}^{*(1)}=[X'G(\lambda^{(1)})X]^{-1}\,X'G(\lambda^{(1)})\underline{Y}=(0.486,\ 0.377)'$$

Up to three decimal places, $\hat{\pi}^{*(1)}$ agrees with $\hat{\pi}^{*(0)}$. Hence the solution of (4.14) is obtained as $\hat{\pi}^{*}=(0.486,\ 0.377)'$. Thus,

$$\hat{\pi}_1=0.486 \qquad \hat{\pi}_2=0.377 \qquad \hat{\pi}_3=1-\hat{\pi}_1-\hat{\pi}_2=0.137$$

TABLE 4.1. Values of the $\{p_{ij}\}$

i \ j	1	2	3
1	$\frac{1}{6}$	$\frac{5}{15}$	$\frac{10}{20}$
2	$\frac{1}{6}$	$\frac{4}{15}$	$\frac{6}{20}$
3	$\frac{1}{6}$	$\frac{3}{15}$	$\frac{3}{20}$
4	$\frac{1}{6}$	$\frac{2}{15}$	$\frac{1}{20}$
5	$\frac{1}{6}$	$\frac{1}{15}$	0
6	$\frac{1}{6}$	0	0

TABLE 4.2

i	1	2	3	4	5	6
$\hat{\lambda}_i$	$\frac{82}{300}$	$\frac{68}{300}$	$\frac{53}{300}$	$\frac{40}{300}$	$\frac{33}{300}$	$\frac{24}{300}$

Now, assuming that 2.5 is a representative variate value for the last class (consisting of people having pre- or extramarital experience with two or more partners), an estimate of μ, the mean number of partners with whom a person in that region had pre- or extramarital sexual relations, is given by

$$\hat{\mu}=0\hat{\pi}_1+1\hat{\pi}_2+2.5\hat{\pi}_3=0.720$$

Warner (1971) describes a general linear model approach, in the context of RR, which is presented briefly below. Let $\underline{Z}$ be a $p\times 1$ random vector with $E(\underline{Z})=\pi$. Some of the elements of $\underline{Z}$ are sensitive measures. Suppose that the problem is to estimate π or some linear function of π without observing the sample values of $\underline{Z}$. The observations on the ith respondent in an SRSWR of size n are represented by the $r\times 1$ vector

$$\underline{Y}_i=T_i\underline{Z}_i$$

where T_i is an observation on an $r\times p$ random matrix of known distribution and $\underline{Z}_i$ represents the $\underline{Z}$ value for the ith respondent. The actual values of T_i and $\underline{Z}_i$ are unknown to the interviewer and the distribution of T_i depends on the specific design adopted. Note that T_i is independent of T_j ($j\neq i$) and also of $\underline{Z}_j$ ($j=1,\ldots,n$). Hence writing

$$E(T_i)=\tau_i$$

one gets $E(\underline{Y}_i)=\tau_i\pi$. Consequently, defining $\underline{Y}=(\underline{Y}'_1,\ldots,\underline{Y}'_n)'$ and $\tau=(\tau'_1,\ldots,\tau'_n)'$, the model can be written as

$$\underline{Y}=\tau\pi+\underline{e}$$

where $E(\underline{e})=0$, disp$(\underline{e})=V$ (say). Thus the linear RR model may be interpreted as an application of the generalized linear regression model, and the method of weighted least squares may be used to estimate π as

$$\hat{\pi}=(\tau'V^{-1}\tau)^{-1}\tau'V^{-1}\underline{Y} \tag{4.18}$$

Warner shows that many of the RR designs discussed earlier are special cases of this model. In practice, if τ is a square matrix, then (4.18) reduces to $\pi=\tau^{-1}\underline{Y}$, and the fact that V is unknown, being dependent on π, poses no problem. However, as in the RR technique considered by Liu *et al.* (1975), the real problem arises when τ is not square. Then one may deal with the unknown V by attempting to solve (4.18) by iterative methods. Sen (1976) suggests the use of two-stage sequential sampling to handle such situations.

4.4. ESTIMATION OF A DISTRIBUTION FUNCTION

The techniques described so far in this chapter deal primarily with the estimation of the mean of a quantitative (sensitive) character. In some situations, however, interest may lie in an estimation of the entire cumulative distribution function (c.d.f.) of such a character. The present section describes the use of the so-called scrambled responses, obtained through a multiplicative or an additive model, in this context. First, the univariate case is considered and then a multivariate extension is indicated.

Poole (1974) considers the problem of estimating the c.d.f. $G(\cdot)$ of a (sensitive) continuous variable X. Denote by $g(\cdot)$ the density of X. Let Y be a random multiplier which is non-negative, continuous, and distributed independently of X. Suppose that Y has a known c.d.f. $R(\cdot)$ and further, let

$$P(Y \leq T) = 1 \tag{4.19}$$

for some known finite T. The RR scheme is such that the respondent generates Y using some specified method and multiplies his or her sensitive answer X by Y. The interviewer receives the scrambled response $Z = XY$. The c.d.f. $F(\cdot)$ of Z is now given by

$$\begin{aligned} F(z) = P(XY \leq z) &= \int_0^T G(zy^{-1})\, dR(y) \\ &= [G(zy^{-1})R(y)]_0^T + \int_0^T zy^{-2} g(zy^{-1})\, R(y)\, dy \\ &= G(zT^{-1}) + \int_{zT^{-1}}^{\infty} g(s) R(zs^{-1})\, ds \qquad -\infty < z < \infty \end{aligned} \tag{4.20}$$

after integration by parts, use of (4.19), and a change in variable. As a special case, if

$$R(y) = \left(\frac{y}{T}\right)^{\alpha+1} \qquad 0 \leq y \leq T \tag{4.21}$$

where $\alpha(> -1)$ is known, then (4.20) reduces to

$$F(z) = G(zT^{-1}) + (zT^{-1})^{\alpha+1} \int_{zT^{-1}}^{\infty} g(s) s^{-(\alpha+1)} ds \tag{4.22}$$

Denote the density corresponding to $F(\cdot)$ by $f(\cdot)$. Then, differentiating (4.22) with respect to z, we obtain

$$f(z)=(\alpha+1)z^{\alpha}T^{-(\alpha+1)}\int_{zT^{-1}}^{\infty} g(s)s^{-(\alpha+1)}ds \tag{4.23}$$

after some simplification. Elimination of the integral between (4.22) and (4.23) yields

$$G(zT^{-1})=F(z)-z(\alpha+1)^{-1}f(z) \tag{4.24}$$

It may be remarked that the generation of Y from (4.21) is easily accomplished since $(Y/T)^{\alpha+1}$ has a uniform distribution over [0,1]. In particular, if $\alpha=0$, $T=1$, then Y itself is uniformly distributed over [0, 1] and (4.24) reduces to the simple form

$$G(z)=F(z)-zf(z) \tag{4.25}$$

If a parametric form for $f(\cdot)$ may be assumed, the parameters of the assumed form may be estimated, and (4.24) or (4.25) may be used in estimating $G(\cdot)$. In general, however, the form of $f(\cdot)$ will not be known, and one will have to employ some nonparametric procedures. Much work has been reported in the literature on the estimation of c.d.f.'s and densities (see, e.g., Parzen, 1962, Leadbetter and Watson, 1963, and Scheult, 1970, among others) and the details in this regard will not be considered here. In particular (see Duffy and Waterton, 1984), defining

$$\begin{aligned} H(z)&=1 \qquad z\geq 0\\ &=0 \qquad z<0 \end{aligned}$$

if $z_1, \ldots, z_n$ are the observed scrambled responses in an SRSWR of size n, then $F(z)$ may be estimated by

$$\hat{F}_n(z)=n^{-1}\textstyle\sum_{i=1}^{n} H(z-z_i) \tag{4.26}$$

while for $f(z)$ one may use the kernel density estimator

$$\hat{f}_n(z)=\frac{\hat{F}_n(z+h)-\hat{F}_n(z-h)}{2h} \tag{4.27}$$

Poole (1974) also describes one useful method of estimation for large samples.

Scrambled responses may as well be obtained by asking the respondent to report $X+Y$, instead of XY, where X is a quantitative sensitive character and Y is a random variable with a known probability distribution. Duffy and Waterton (1984) consider some details of such an additive model treating the discrete and continuous cases separately.

First suppose that the sensitive character X is discrete, assuming nonnegative integral values with a c.d.f. $G(\cdot)$ and a probability mass function (p.m.f.) $g(\cdot)$. Let Y follow the geometric distribution

$$P(Y=y)=\pi(1-\pi)^y \qquad (y=0,1,2,\ldots)$$

where π is known. The respondent is asked to conduct a sequence of Bernoullian trials each with a probability of success π, count the number of failures preceding the first success to generate Y, and then report the scrambled response $Z=X+Y$. Denoting the c.d.f. and the p.m.f. of Z by $F(\cdot)$ and $f(\cdot)$, respectively, one now obtains

$$G(z)=F(z)+\frac{(1-\pi)f(z)}{\pi} \qquad (z=0,1,2,\ldots) \tag{4.28}$$

To prove the validity of (4.28), note that

$$f(r)=P(X+Y=r)=\textstyle\sum_{i=0}^{r} g(i)(1-\pi)^{r-i}\pi \qquad (r=0,1,2,\ldots)$$

since X and Y are independent. Hence

$$g(r)=\pi^{-1}[f(r)-(1-\pi)f(r-1)] \qquad (r=0,1,2,\ldots) \tag{4.29}$$

where $f(-1)=0$. Summing (4.29) over $r=0, 1, \ldots, z$, the relation (4.28) follows immediately.

With an SRSWR of size n, Duffy and Waterton (1984) suggest a UE of $G(z)$, based on (4.28), as

$$\hat{G}_n(z)=\hat{F}_n(z)+\frac{(1-\pi)\hat{f}_n(z)}{\pi} \tag{4.30}$$

where $\hat{F}_n(z)$ is given by (4.26) and

$$\hat{f}_n(z)=n^{-1}\textstyle\sum_{i=1}^{n} \Delta(z-z_i)$$

with

$$\begin{aligned}\Delta(z)&=1 \qquad z=0\\ &=0, \qquad z\neq 0\end{aligned}$$

By well-known results on multinomial proportions,

$$\begin{aligned}
\text{var}[\hat{F}_n(z)] &= n^{-1}F(z)[1-F(z)] \\
\text{var}[\hat{f}_n(z)] &= n^{-1}f(z)[1-f(z)] \\
\text{cov}[\hat{F}_n(z), \hat{f}_n(z)] &= \text{cov}[\hat{f}_n(z) + \hat{F}_n(z-1), \hat{f}_n(z)] \\
&= n^{-1}\{f(z)[1-f(z)] - F(z-1)f(z)\} \\
&= n^{-1}f(z)[1-F(z)]
\end{aligned}$$

and from (4.30) it follows that

$$\text{var}[\hat{G}_n(z)] = n^{-1}\left\{F(z)[1-F(z)] + \left(\frac{1-\pi}{\pi}\right)^2 f(z)[1-f(z)] + 2\left(\frac{1-\pi}{\pi}\right) f(z)[1-F(z)]\right\} \tag{4.31}$$

Considering now the additive model for a continuous (sensitive) variable X, suppose that X is nonnegative with a c.d.f. $G(\cdot)$, and let the scrambling component Y be distributed exponentially with density $\alpha e^{-\alpha y}(y \geq 0)$, where α (>0) is known. As before, the respondent generates Y and reports only $Z = X + Y$. Denote the c.d.f. and the density of Z by $F(\cdot)$ and $f(\cdot)$, respectively. Then

$$\begin{aligned}
F(z) = P(X + Y \leq z) &= \int_0^z P(Y \leq z - x)\, \mathrm{d}G(x) \\
&= \int_0^z [1 - e^{-\alpha(z-x)}]\mathrm{d}G(x) \\
&= G(z) - e^{-\alpha z}\int_0^z e^{\alpha x}\, \mathrm{d}G(x)
\end{aligned} \tag{4.32}$$

Differentiating (4.32) with respect to z yields

$$f(z) = \alpha e^{-\alpha z}\int_0^z e^{\alpha x}\, \mathrm{d}G(x) \tag{4.33}$$

and hence

$$G(z) = F(z) + \alpha^{-1}f(z) \qquad (z \geq 0)$$

In view of the above, Duffy and Waterton (1984) suggest an estimator of $G(z)$, based on the scrambed responses available from an SRSWR of size n, as

$$\hat{G}_n(z) = \hat{F}_n(z) + \alpha^{-1}\hat{f}_n(z) \tag{4.34}$$

where $\hat{F}_n(z)$ and $\hat{f}_n(z)$ are as in (4.26) and (4.27), respectively. Since

$$E[\hat{f}_n(z)] = \frac{F(z+h) - F(z-h)}{2h} \neq f(z)$$

$\hat{G}_n(z)$ is biased as an estimator of $G(z)$, and the amount of bias is given by

$$\begin{aligned}\text{bias}[\hat{G}_n(z)] &= E[\hat{G}_n(z)] - G(z)\\ &= \alpha^{-1}\{(2h)^{-1}[F(z+h) - F(z-h)] - f(z)\}\end{aligned}$$

To find the variance of $\hat{G}_n(z)$, note that by (4.27) and (4.34),

$$\hat{G}_n(z) = U_1 + a_2 U_2 + a_3 U_3$$

where

$$U_1 = \hat{F}_n(z-h) \qquad U_2 = \hat{F}_n(z) - \hat{F}_n(z-h) \qquad U_3 = \hat{F}_n(z+h) - \hat{F}_n(z)$$
$$a_2 = 1 + (2\alpha h)^{-1} \qquad a_3 = (2\alpha h)^{-1}$$

Clearly, U_1, U_2, and U_3 are sample multinomial proportions corresponding to mutually exclusive classes, the corresponding true proportions being given by

$$\xi_1 = F(z-h) \qquad \xi_2 = F(z) - F(z-h) \qquad \xi_3 = F(z+h) - F(z)$$

respectively. Hence

$$\begin{aligned}\text{var}[\hat{G}_n(z)] &= n^{-1}[\xi_1(1-\xi_1) + a_2^2\,\xi_2(1-\xi_2) + a_3^2\xi_3(1-\xi_3)\\ &\quad - 2a_2\xi_1\xi_2 - 2a_3\xi_1\xi_3 - 2a_2a_3\xi_2\xi_3]\\ &= n^{-1}[\xi_1 + a_2^2\xi_2 + a_3^2\xi_3 - (\xi_1 + a_2\xi_2 + a_3\xi_3)^2]\end{aligned}$$

The consistency of $\hat{G}_n(z)$ as $h \to 0$ and $nh \to \infty$ follows from that of $\hat{f}_n(z)$ (see, e.g., Rosenblatt, 1956).

Under both multiplicative and additive models, it is fairly straightforward to obtain UEs of μ_X, the population mean of the sensitive character X, which may be discrete or continuous. As before, let Y be the scrambling component, distributed independently of X, with a known probability distribution and Z be the scrambled response. Also, let

$$\mu_Y = E(Y) \qquad \mu_z = E(Z)$$
$$\sigma_X^2 = \text{var}(X) \qquad \sigma_Y^2 = \text{var}(Y) \qquad \sigma_Z^2 = \text{var}(Z)$$

Under the multiplicative model, $Z = XY$ and $\mu_Z = \mu_X \mu_Y$. Hence if $z_1, \ldots,$ z_n be the observed scrambled responses in an SRSWR of size n and

$\bar{z}=n^{-1}\Sigma_{i=1}^{n} z_i$, then a UE of μ_X is obtained as $\hat{\mu}_X=\bar{z}/\mu_Y$, with

$$\begin{aligned}\operatorname{var}(\hat{\mu}_X) &= \frac{\sigma_Z^2}{n\mu_Y^2} \\ &= \frac{\sigma_X^2\sigma_Y^2+\sigma_X^2\mu_Y^2+\mu_X^2\sigma_Y^2}{n\mu_Y^2}\end{aligned} \tag{4.35}$$

For an additive model $Z=X+Y$, so that $\mu_Z=\mu_X+\mu_Y$ and $\hat{\mu}_X=\bar{z}-\mu_Y$ is a UE of μ_X with

$$\operatorname{var}(\hat{\mu}_X')=\frac{\sigma_Z^2}{n}=\frac{\sigma_X^2+\sigma_Y^2}{n} \tag{4.36}$$

Under either model it is easy to obtain a UE of $\operatorname{var}(\hat{\mu}_X)$ replacing σ_Z^2 in (4.35) or (4.36) by $s_z^2=(n-1)^{-1}\,\Sigma_{i=1}^{n}(z_i-\bar{z})^2$. It may be remarked that in so far as the estimation of μ_X is concerned, one need not know the entire distribution of Y, and knowledge of μ_Y alone serves the purpose. Therefore, instead of making explicit use of a chance mechanism to generate Y, one may as well take Y as an unrelated innocuous character with a known population mean, but then in order that the multiplicative model be valid, Y should be statistically independent of the study variable X. For further details regarding the estimation of μ_X under the multiplicative and additive models, we refer to Pollock and Bek (1976) and Eichhorn and Hayre (1983).

We conclude this section with a brief discussion on the problem of estimating a c.d.f. in a multivariate setup. Poole and Clayton (1982) extend the results in Poole (1974) to a sensitive k-dimensional continuous random variable $\underset{\sim}{X}=(X_1,\ldots,X_k)'$ in the k-dimensional Euclidean space. The ranges of the components of $\underset{\sim}{X}$ are assumed to be functionally independent. A respondent is asked to report the scrambled vector response

$$\underset{\sim}{Z}=(Z_1,\ldots,Z_k)'=(X_1Y_1,\ldots,X_kY_k)'$$

where Y_i, the scrambling component for X_i, is an observation from the distribution

$$P(Y_i\le y_i)=\left(\frac{y_i}{T_i}\right)^{\alpha_i+1} \qquad 0\le y_i\le T_i \qquad (i=1,\ldots,k) \tag{4.37}$$

$\alpha_i>-1$ $(i=1,\ldots,k)$, $Y_1,\ldots,Y_k$ are independent and $\alpha_1,\ldots,\alpha_k$, $T_1,\ldots,T_k$ are known. If $F(\cdot)$ and $G(\cdot)$ are the c.d.f.'s of $\underset{\sim}{Z}$ and $\underset{\sim}{X}$, respectively, it can be shown that

$$G\left(\frac{z_1}{T_1},\ldots,\frac{z_k}{T_k}\right)=\left\{\prod_{i=1}^{k}\left[1-\frac{z_i\partial}{(\alpha_i+1)\partial z_i}\right]\right\}F(z_1,\ldots,z_k) \tag{4.38}$$

In (4.38), the expression in braces is the linear differential operator given by

$$1-\frac{z_1\partial}{(\alpha_1+1)\partial z_1} \qquad \text{for } k=1$$

$$1-\frac{z_1\partial}{(\alpha_1+1)\partial z_1}-\frac{z_2\partial}{(\alpha_2+1)\partial z_2}+\frac{z_1 z_2}{(\alpha_1+1)(\alpha_2+1)}\quad\frac{\partial^2}{\partial z_1\partial z_2} \qquad \text{for } k=2$$

and so on. The validity of (4.38) for $k=1$ is immediate from (4.24). The proof of (4.38) for general k, which involves induction on k, is available in Poole and Clayton (1982) and left here as an exercise (Exercise 4.4).

As Poole and Clayton (1982) indicate, the relation (4.38) may be used to show that under the additive multivariate model (i.e., each component is scrambled by adding a random number),

$$G(z_1-T_1,\ldots,z_k-T_k)=\left[\prod_{i=1}^{k}\left(1-\frac{\partial}{\beta_i\partial z_i}\right)\right]F(z_1,\ldots,z_k) \quad (4.39)$$

where, as before, $G(\cdot)$ is the c.d.f. of a sensitive vector of random variables $\underset{\sim}{X}=(X_1,\ldots,X_k)'$, $F(\cdot)$ is the c.d.f. of the scrambled response vector $\underset{\sim}{Z}=(Z_1,\ldots,Z_k)'$, and the scrambling distribution function for the ith component is

$$R_i(y_i)=e^{\beta_i(y_i-T_i)} \qquad y_i\le T_i<\infty \qquad \beta_i>0 \qquad (i=1,\ldots,k) \quad (4.40)$$

This follows from (4.38) and the fact that if $Z_i=X_i+Y_i$ with Y_i distributed as in (4.40); then $e^{Z_i}=e^{X_i}e^{Y_i}$ and e^{Y_i} is distributed in the form (4.37).

In estimating $G(\cdot)$ from (4.38) or (4.39), one has to estimate $F(\cdot)$ in such a way as to provide estimates of all mixed partial derivatives up to order k. Poole and Clayton (1982) indicate the use of nonparametric methods in this context.

4.5. APPLICATIONS OF HOEFFDING'S U STATISTIC AND VON MISES' DIFFERENTIABLE STATISTICAL FUNCTIONS

Sen (1974) considers the problem of optimal unbiased estimation of regular functionals of cumulative distribution functions (c.d.f.'s) and extends to RR the distribution theory of Hoeffding's U statistic and Von Mises' differentiable statistical functions.

Let $\underset{\sim}{X}$ be a k-component vector of sensitive variables and $\underset{\sim}{Y}$ a k-component vector of unrelated nonsensitive variables. As in Section 4.2, suppose that two independent SRSWRs of sizes n_1 and n_2 are drawn from the population. For $i=1$, 2, a respondent in the ith sample reports on $\underset{\sim}{X}$

with probability P_i and on $\underset{\sim}{Y}$ with Q_i ($=1-P_i$), and let the observations in the ith sample be $\underset{\sim}{Z}_{i1}, \ldots, \underset{\sim}{Z}_{in_i}$, which are independent and identically distributed with a common c.d.f. $F_i(\cdot)$.

If the responses on $\underset{\sim}{X}$ and $\underset{\sim}{Y}$ have c.d.f.'s $G(\cdot)$ and $H(\cdot)$, then

$$F_i(\underset{\sim}{x}) = P_i G(\underset{\sim}{x}) + Q_i H(\underset{\sim}{x}) \qquad (i=1,2) \tag{4.41}$$

where $\underset{\sim}{x} \in \mathscr{R}^k$, the k-dimensional Euclidean space. Assuming that $P_1 > P_2$, the above yields

$$G(\underset{\sim}{x}) = \frac{Q_2 F_1(\underset{\sim}{x}) - Q_1 F_2(\underset{\sim}{x})}{(P_1 - P_2)}$$

$$H(\underset{\sim}{x}) = \frac{P_1 F_2(\underset{\sim}{x}) - P_2 F_1(\underset{\sim}{x})}{P_1 - P_2}$$

Thus although $F_1(\cdot)$ and $F_2(\cdot)$ are convex combinations of $G(\cdot)$ and $H(\cdot)$, the converse is not true.

Let $\mathscr{F}$ be a class c.d.f.'s containing all convex combinations of $G(\cdot)$ and $H(\cdot)$, and for any $F(\cdot)$ in $\mathscr{F}$, let $\theta(F)$ be a regular functional on $\mathscr{F}$ of degree m and having a symmetric kernel $\phi(\xi_1, \ldots, \xi_m)$, that is,

$$\theta(F) = \int_{\mathscr{R}^{km}} \cdots \int \phi(\xi_1, \ldots, \xi_m)\, dF(\xi_1) \cdots dF(\xi_m) \qquad \forall\, F \in \mathscr{F}$$

Here ξ_i's are unobservable k vectors and one observes $\underset{\sim}{Z}_{ij}$'s from $F_i(\cdot)$ $(i=1,2)$. The problem is to estimate $\theta(G)$. Sen (1974) shows that for RR as above, a regular functional $\theta(G)$ of degree m (≥ 1) is estimable if n_1, $n_2 \geq m$.

To find a uniformly minimum variance unbiased estimator (UMVUE) of $\theta(G)$, write for $s=0, 1, \ldots, m$,

$$\theta_s(F_1, F_2) = \int_{\mathscr{R}^{km}} \cdots \int \phi(\underset{\sim}{Z}_1, \ldots, \underset{\sim}{Z}_m) \prod_{j=1}^{m-s} dF_1(\underset{\sim}{Z}_j) \prod_{t=m-s+1}^{m} dF_2(\underset{\sim}{Z}_t)$$

and observe that corresponding to $\theta_s(F_1, F_2)$ a generalized U statistic is

$$U_s(n_1, n_2) = \frac{1}{\binom{n_1}{m-s}\binom{n_2}{s}} \sum_s{}^{*} \phi(\underset{\sim}{Z}_{1i_1}, \ldots, \underset{\sim}{Z}_{1i_{m-s}}, \underset{\sim}{Z}_{2j_1}, \ldots, \underset{\sim}{Z}_{2j_s}) \tag{4.42}$$

where Σ_s^* is a sum over all choices of $(i_1, \ldots, i_{m-s})$ and $(j_1, \ldots, j_s)$ such that $1 \leq i_1 < \cdots < i_{m-s} \leq n_1$, $1 \leq j_1 < \cdots < j_s \leq n_2$. This $U_s(n_1, n_2)$, being symmetric in $Z_{11}, \ldots, Z_{1n_1}$ and also in $Z_{21}, \ldots, Z_{2n_2}$, is a function of the two-sample order statistics. Since these order statistics are complete, $U_s(n_1, n_2)$ is the UMVUE of $\theta_s(F_1, F_2)$, $s=0, 1, \ldots, m$. In fact, the vector

$$[U_0(n_1, n_2), \ldots, U_m(n_1, n_2)]'$$

is the "minimum concentration ellipsoid" unbiased estimator of

$$[\theta_0(F_1, F_2), \ldots, \theta_m(F_1, F_2)]'$$

Hence recalling (4.41), one gets

$$\theta(G) = \frac{1}{(P_1 - P_2)^m} \sum_{s=0}^{m} \binom{m}{s} (-1)^s Q_2^{m-s} Q_1^s \theta_s(F_1, F_2)$$

and it follows that the UMVUE of $\theta(G)$ is given by

$$U(n_1, n_2) = \frac{}{(P_1 - P_2)^m} \sum_{s=0}^{m} \binom{m}{s} (-1)^s Q_2^{m-s} Q_1^s U_s(n_1, n_2) \qquad (4.43)$$

Example 4.3. Let $\mathscr{F} = \{F \mid \int_{-\infty}^{\infty} x^2 \mathrm{d}F < \infty\}$. Then $\mathscr{F}$ is convex. Consider

$$\theta(F) = \int_{-\infty}^{\infty} x^2 \mathrm{d}F - \left(\int_{-\infty}^{\infty} x \mathrm{d}F \right)^2 = \sigma^2(F)$$

say, which is a regular functional of degree $m=2$ on $\mathscr{F}$ having a symmetric kernel

$$\phi(\xi_1, \xi_2) = \frac{1}{2}(\xi_1 - \xi_2)^2$$

Hence, by (4.42), observing that

$$U_0(n_1, n_2) = \frac{1}{n_1 - 1} \sum_{j=1}^{n_1} (Z_{1j} - \bar{Z}_1)^2$$

$$U_1(n_1, n_2) = \frac{1}{n_1 n_2} \sum_{j=1}^{n_1} \sum_{t=1}^{n_2} (Z_{1j} - Z_{2t})^2$$

$$U_2(n_1, n_2) = \frac{1}{n_2 - 1} \sum_{j=1}^{n_2} (Z_{2j} - \bar{Z}_2)^2$$

where $\bar{Z}_i = n_i^{-1} \Sigma_{j=1}^{n_i} Z_{ij} (i=1,2)$, one may apply (4.43) to obtain the UMVUE of $\sigma^2(G)$, the variance of the sensitive character, as

$$U(n_1, n_2) = \frac{1}{(P_1 - P_2)^2} [Q_2^2 U_0(n_1, n_2) - 2Q_1 Q_2 U_1(n_1, n_2) + Q_1^2 U_2(n_1, n_2)]$$

In an alternative approach, employing Von Mises' differentiable statistical functions, consider the empirical c.d.f.'s

$$F_{n_i}(\underset{\sim}{x}) = n_i^{-1} \sum_{j=1}^{n_i} H(\underset{\sim}{x} - \underset{\sim}{Z}_{ij}) \qquad \underset{\sim}{x} \in \mathscr{R}^k \qquad (i=1,2)$$

[where $H(\underset{\sim}{x})=1$ if $\underset{\sim}{x} \geq \underset{\sim}{0}; =0$ otherwise] which are sufficient, but not necessarily minimal sufficient, for $F_1(\cdot)$ and $F_2(\cdot)$. Von Mises' generalized differentiable statistical function for $\theta_s(F_1, F_2)$ $(s=0, 1, \ldots, m)$ is

$$V_s(n_1, n_2) = \int_{\mathscr{R}^{km}} \cdots \int \phi(\xi_1, \ldots, \xi_m) \prod_{j=1}^{m-s} \mathrm{d}F_{n_1}(\xi_j) \prod_{t=m-s+1}^{m} \mathrm{d}F_{n_2}(\xi_t)$$

$$= \frac{1}{n_1^{m-s} n_2^s} \sum_{i_1=1}^{n_1} \cdots \sum_{i_{m-s}=1}^{n_1} \sum_{j_1=1}^{n_2} \cdots \sum_{j_s=1}^{n_2} \phi(\underset{\sim}{Z}_{1i_1}, \ldots, \underset{\sim}{Z}_{1i_{m-s}}, \underset{\sim}{Z}_{2j_1}, \ldots, \underset{\sim}{Z}_{2j_s})$$

In analogy with (4.43), this gives an alternative estimator of $\theta(G)$ as

$$V(n_1, n_2) = \frac{1}{(P_1 - P_2)^m} \sum_{s=0}^{m} \binom{m}{s} (-1)^s Q_2^{m-s} Q_1^s \, V_s(n_1, n_2)$$

It may be noted that $V(n_1, n_2)$, unlike $U(n_1, n_2)$, is not in general unbiased for $\theta(G)$. As $n \, (=(n_1+n_2)) \to \infty$, n_1/n_2 being bounded away from 0, 1, it can be shown that

$$\sqrt{\mathrm{n}}[U(n_1, n_2) - V(n_1, n_2)]$$

converges in probability to zero.

Sen (1974) also gives expressions for $\mathrm{var}[U(n_1, n_2)]$ and shows that

$$\sqrt{n}[U(n_1, n_2) - \theta(G)]$$

converges in law to the normal distribution with mean 0 and variance

$$\sigma^2 = \frac{m^2}{(P_1 - P_2)^{2m}} \sum_{s=0}^{m} \sum_{s'=0}^{m} (-1)^{s+s'} Q_1^{s+s'} Q_2^{2m-s-s'} \times \left[\frac{1}{\lambda} \binom{m-1}{s} \binom{m-1}{s'} S_{10}(s, s'; G, H) + \frac{1}{1-\lambda} \binom{m-1}{s-1} \binom{m-1}{s'-1} S_{01}(s, s'; G, H) \right]$$

where $\lambda = \lim(n_1/n)$ as $n \to \infty$, and S_{10} and S_{01} are regular (and hence estimable) functions of $G(\cdot)$ and $H(\cdot)$.

A consistent estimator $\hat{\sigma}^2_{n_1,n_2}$ for σ^2 is available from Sen (1960) and by Slutsky's theorem,

$$\frac{\sqrt{n}[U(n_1,n_2)-\theta(G)]}{\hat{\sigma}_{n_1,n_2}}$$

is asymptotically normal (0, 1). This may be used in getting large-sample test and confidence limits for $\theta(G)$.

EXERCISES

4.1. (Chow and Liu, 1973) Even in dealing with a qualitative character, essentially quantitative responses may be of use. Consider a dichotomous population with respect to a sensitive attribute A, the true proportion of A being π_A. To construct a randomization device T balls of colors red and blue are put in proportions $P:(1-P)$ in a bottle with a long transparent neck on which at most $t(<T)$ balls can stand one after another. After shaking the bottle thoroughly, each respondent in an SRSWR of size n reports the number of red (blue) balls in the neck if he or she belongs to A($\bar{\text{A}}$). Let $\bar{Z}$ stand for the sample mean response. Show that

$$E(\bar{Z})=t[P\pi_A+(1-P)(1-\pi_A)]$$

Hence suggest a UE of π_A. Derive the variance of your estimator and suggest a UE of this variance.

4.2. (Greenberg *et al.*, 1971) Under the setup of Section 4.2, suppose that X and Y are both normally distributed with respective means μ_X and μ_Y and variances σ^2_X and σ^2_Y. Then show that

$$\mathrm{E}\left\{\left[\frac{\mathrm{d}}{\mathrm{d}\mu_X}\log f_i(Z)\right]^2\right\}$$

$$=\frac{P_i}{\sqrt{2\pi\sigma^2_X}}\int_{-\infty}^{\infty}\frac{t^2\exp[-(1/2)t^2]}{1+[(1-P_i)/P_i]\phi_1\exp[(1/2)t^2-(1/2)(\phi_1 t+\phi_2)^2]}\mathrm{d}t \qquad (i=1,2)$$

where $\phi_1=\sigma_X/\sigma_Y$ and $\phi_2=(\mu_X-\mu_Y)/\sigma_Y$

Hence taking $n_1=n_2$, examine how one can compute numerically the efficiency of the estimator $\hat{\mu}_X$ given by (4.6), with respect to the lower bound (4.8), for different values of P_1, P_2, ϕ_1 and ϕ_2.

4.3. Prove (4.10).
4.4. Prove (4.38) by induction on k. Also examine in detail the derivation of (4.39) from (4.38).

REFERENCES

Chow, L. P., and Liu, P. T. (1973). A new RR technique: the multiple answer model. Department of Population Dynamics, Johns Hopkins University, Baltimore, Md.

Dalenius, T., and Vitale, R. A. (1974). A new RR design for estimating the mean of a distribution. *Tech. Rep.* 78, Brown University, Providence, R. J.

Duffy, J. C., and Waterton, J. J. (1984). RR models for estimating the distribution function of a quantitative character. *Internat. Statist. Rev.* **52**, 165–171.

Eichhorn, B. H., and Hayre, L. S. (1983). Scrambled RR models for obtaining sensitive quantitative data. *J. Statist. Plann. Inference* **7**, 307–316.

Greenberg, B. G., Kubler, R. R., Abernathy, J. R., and Horvitz, D. G. (1971). Applications of the RR technique in obtaining quantitative data. *J. Amer. Statist. Assoc.* **66** 243–250.

Leadbetter, M. R., and Watson, G. S. (1963). On the estimation of the probability density I. *Ann. Math. Statist.* **34**, 480–491.

Liu, P. T., Chow, L. P., and Mosley, W. H. (1975) Use of the RR technique with a new randomizing device. *J. Amer. Statist. Assoc.* **70**, 329–332.

Parzen, E. (1962). On estimation of a probability density and mode. *Ann. Math. Statist.* **33**, 1065–1076.

Pollock, K. H., and Bek, Y. (1976). A comparison of three RR models for quantitative data. *J. Amer. Statist. Assoc.* **71**, 884–886.

Poole, W. K. (1974). Estimation of the distribution function of a continuous type random variable through RR. *J. Amer. Statist. Assoc.* **69**, 1002–1005.

Poole, W. K., and Clayton, A. C. (1982). Generalizations of a contamination model for continuous type random variables. *Comm. Statist. Theory Methods* **11**, 1733–1742.

Rao, C. R. (1973). *Linear Statistical Inference and Its Applications*, 2nd ed. John Wiley, New York.

Rosenblatt, M. (1956). Remarks on some nonparametric estimates of a density function. *Ann. Math. Statist.* **27**, 832–835.

Scheult, A. H. (1970). On unbiased estimation of density functions. Ph.D. thesis, North Carolina State University.

Sen, P. K. (1960). On some convergence properties of U-statistics. *Bull. Calcutta Statist. Assoc.* **10**, 1–18.

Sen, P. K. (1974). On unbiased estimation for RR models. *J. Amer. Statist. Assoc.* **69**, 997–1001.

Sen, P. K. (1976). Asymptotically optimal estimators of general parameters in RR models. *Internat. Statist. Rev.* **44**, 223–224.

Warner, S. L. (1971). The linear RR model. *J. Amer. Statist. Assoc.* **66**, 884–888.

5

Efficient Estimation and Protection of Privacy

5.1. INTRODUCTION

Except when stated otherwise, we assume truthful reporting in randomized queries. In RR surveys, although an individual respondent is not asked to divulge his or her true standing in respect to a sensitive characteristic, the person does, nevertheless, run certain risks of disclosures. It is possible that among the respondents there may be intelligent and knowledgeable people well equipped intellectually to analyze and weigh the hazards in giving out secrets. Naturally, they must be convinced that their privacy is well guarded before they will be persuaded to make available damaging and incriminating documents. So protection of confidentiality is an important ingredient of RR theory and practice and we should dwell on this issue.

In the context of the Warner model introduced in Chapter 1, it was observed that maintenance of privacy and efficient estimation with RR were in conflict. In this chapter this will be seen to be a general phenomenon true for other models as well. So it behooves us to treat a problem of optimally efficient estimation in RR subject to practical constraints imposed by the requirement of protecting the privacy of a respondent. Discussions based on findings and recommendations of many of our colleagues in the field of RR theory and practice follow, somewhat in brief.

5.2. DICHOTOMOUS POPULATION: "YES-NO" RESPONSE

Consider a population divided into complementary sensitive groups A and $\bar{A}$ with unknown proportions π_A and $1-\pi_A$, respectively. Considering a dichotomous response model, a typical response is R, which is "yes" (say, y)

or "no" (say, $\bar{y}$). The conditional probabilities that a response R comes from an individual of groups A and Ā, respectively, are $P(R|A)$ and $P(R|\bar{A})$. These are quantities at the investigator's disposal.

The posterior probabilities that a respondent belongs to groups A and Ā, respectively, when he or she reports R are, say, $P(A|R)$ and $P(\bar{A}|R)$. These are the revealing probabilities (Lanke, 1975; Anderson, 1975 a, b, c). By Bayes' theorem,

$$P(A|R) = \frac{\pi_A P(R|A)}{D} \qquad P(\bar{A}|R) = 1 - P(A|R)$$

where

$$D = \pi_A P(R|A) + (1 - \pi_A) P(R|\bar{A}) \tag{5.1}$$

Following Leysieffer and Warner (1976), the response R is regarded as jeopardizing with respect to A or Ā if

$$P(A|R) > \pi_A \qquad \text{or} \qquad P(\bar{A}|R) > 1 - \pi_A \tag{5.2}$$

respectively. Since

$$\frac{P(A|R)}{P(\bar{A}|R)} \; \frac{1-\pi_A}{\pi_A} = \frac{P(R|A)}{P(R|\bar{A})} \tag{5.3}$$

it follows that if the right-hand member of (5.3) is greater (less) than unity, then R is jeopardizing with respect to A ($\bar{A}$) in the sense that with this response a respondent genuinely of group A ($\bar{A}$) yields an increased posterior probability that he or she is in group A ($\bar{A}$) rather than in group $\bar{A}$ (A) and thus tilts the scale against himself or herself if A ($\bar{A}$) is stigmatizing.

Consideration of relations (5.2) and (5.3) led Leysieffer and Warner (1976) to propose the following natural measures of jeopardy carried by R about A and $\bar{A}$, respectively:

$$g(R|A) = \frac{P(R|A)}{P(R|\bar{A})} \qquad g(R|\bar{A}) = \frac{1}{g(R|A)} \tag{5.4}$$

The response R is nonjeopardizing if and only if

$$g(R|A) = 1 \tag{5.5}$$

Now, clearly, the probability of a "yes" response is

$$\begin{aligned} \lambda &= P(y|A)\pi_A + P(y|\bar{A})\,(1-\pi_A) \\ &= [P(y|A) - P(y|\bar{A})]\pi_A + P(y|\bar{A}) \end{aligned} \tag{5.6}$$

If an SRSWR of size n is taken and $\hat{\lambda}$ is the sample proportion of "yes" replies, then by (5.6), a UE of π_A is

$$\hat{\pi}_A = \frac{\hat{\lambda} - P(y|\bar{A})}{P(y|A) - P(y|\bar{A})}$$

which is defined if and only if

$$P(y|A) - P(y|\bar{A}) \neq 0 \tag{5.7}$$

that is, if and only if (5.5) is violated. In fact, because of the completeness (Lehmann, 1959) of $\hat{\lambda}$, existence of a UE for π_A necessarily makes a response jeopardic with respect to either A or $\bar{A}$.

Now suppose that the RR procedure is designed so that (5.7) is satisfied allowing unbiased estimation of π_A even at the cost of privacy, which has to be sacrificed to some extent. Let us see how it affects the efficiency. It is easy to work out the variance formula for $\hat{\pi}_A$ as

$$\begin{aligned}\mathrm{var}(\hat{\pi}_A) &= \frac{\pi_A(1-\pi_A)}{n} \\ &\quad + \frac{\pi_A P(y|A)\,[1-P(y|A)] + (1-\pi_A)P(y|\bar{A})\,[1-P(y|\bar{A})]}{n[P(y|A)-P(y|\bar{A})]^2} \\ &= \frac{\pi_A(1-\pi_A)}{n} + \frac{\pi_A g(y|A) + (1-\pi_A)g(\bar{y}|\bar{A})}{n\{[g(y|A)-1]\,[g(\bar{y}|\bar{A})-1]\}}.\end{aligned} \tag{5.8}$$

Assuming without loss of generality that $P(y|A) > P(y|\bar{A})$, so that

$$g(y|A) > 1 \qquad g(\bar{y}|\bar{A}) > 1$$

(and hence y and $\bar{y}$ are jeopardizing for A and $\bar{A}$, respectively), it follows from (5.8) that

$$\frac{\partial\,\mathrm{var}(\hat{\pi}_A)}{\partial g(y|A)} < 0 \qquad \frac{\partial\,\mathrm{var}(\hat{\pi}_A)}{\partial g(\bar{y}|\bar{A})} < 0$$

Hence for the sake of efficiency, one needs as large magnitudes as possible for $g(y|A)$ and $g(\bar{y}|\bar{A})$, both above unity.

From practical considerations regarding protection of privacy, one can fix some maximal allowable levels of $g(y|A)$ and $g(\bar{y}|\bar{A})$, say k_1 and k_2, respectively. Thus the problem now becomes one of constrained optimization, that is, of minimizing var $(\hat{\pi}_A)$ subject to

$$1 < g(y|A) < k_1 \qquad 1 < g(\bar{y}|\bar{A}) < k_2 \tag{5.9}$$

A solution to this optimization problem may be obtained by choosing the design parameters such that $g(y|A)$ and $g(\bar{y}|\bar{A})$ are as large as possible subject to (5.9).

Essentially similar analysis on this problem of efficient estimation protecting privacy was carried out by Anderson (1975a, b, c; 1976; 1977), who reached parallel conclusions. Anderson (1975b) calls $P(A|R)$ and $P(\bar{A}|R)$ the two "risks of suspicion" corresponding to response R. He suggests restricting them such that

$$P(A|R) \leq \xi_2 < 1 \qquad P(\bar{A}|R) \leq 1 - \xi_1 < 1 \tag{5.10}$$

respectively, if A and $\bar{A}$ are embarrassing, suitably fixing ξ_1 and ξ_2 in (0, 1). Evidently, (5.10) demands fixing the design parameters such that

$$\xi_1 \leq P(A|R) \leq \xi_2 \tag{5.11}$$

Of course, $P(A|R)$ depends on π_A, which is in [0,1]. Naturally, a relation (5.11) may be satisfied only for values of π_A in a subinterval of [0,1], say $[\pi(1), \pi(2)]$. Hence by (5.1), (5.4), and (5.11), for every π_A in $[\pi(1), \pi(2)]$,

$$g(y|A) \leq \frac{1-\pi_A}{\pi_A} \frac{\xi_2}{1-\xi_2}$$

$$g(\bar{y}|\bar{A}) \leq \frac{\pi_A}{1-\pi_A} \frac{1-\xi_1}{\xi_1}.$$

This implies that

$$g(y|A) \leq \frac{1-\pi(2)}{\pi(2)} \frac{\xi_2}{1-\xi_2}$$

$$g(\bar{y}|\bar{A}) \leq \frac{\pi(1)}{1-\pi(1)} \frac{1-\xi_1}{\xi_1}$$

Thus Anderson's (1975a, b, c) criterion for protection of confidentiality effectively sets upper bounds for $g(y|A)$ and $g(\bar{y}|\bar{A})$ and for minimizing $\text{var}(\hat{\pi}_A)$, subject to the foregoing restrictions, it is enough to take $g(y|A)$ and $g(\bar{y}|\bar{A})$, with this approach as well, at their maximal allowable levels.

Consider now, in particular, the Warner model of Section 1.2 and suppose that a respondent reports his or her status relative to A with probability P and to $\bar{A}$ with $1-P$. Then clearly

$$P(y|A) = P(\bar{y}|\bar{A}) = P \qquad P(\bar{y}|A) = P(y|\bar{A}) = 1-P$$

and $P(y|A) > P(y|\bar{A})$ if $P > 1/2$. Now from (5.4), the jeopardy function is given by, say,

$$g(y|A) = g_W(y|A) = \frac{P}{1-P} \qquad g(\bar{y}|\bar{A}) = g_W(\bar{y}|\bar{A}) = \frac{P}{1-P}$$

Thus if $k_1 = k_2 = k$, say, maximization of $g_W(y|A)$ and $g_W(\bar{y}|\bar{A})$ leads to a design with $P/(1-P) = k$ [i.e., $P = k/(k+1)$]. If, however, $k_1 \neq k_2$, as $g_W(y|A) = g_W(\bar{y}|\bar{A})$, different upper bounds for them cannot be attained simultaneously. In that case if, without loss of generality, $k_1 < k_2$, one should so design that $P/(1-P) = k_1$, that is,

$$P = \frac{k_1}{k_1 + 1} \tag{5.12}$$

Next consdier the unrelated question model of Section 2.2. Suppose that π_Y, the true proportion corresponding to the unrelated character Y, is known. Further, let P_1 and $1 - P_1$ be the respective probabilities that the response relates to the attributes A and Y, this slight change in the notation from Section 2.2 being imperative in order to compare this model with the Warner model. Then assuming the attributes A and Y to be independent,

$$P(y|A) = P_1 + (1-P_1)\pi_Y \qquad P(y|\bar{A}) = (1-P_1)\pi_Y$$

$$P(\bar{y}|A) = (1-P_1)(1-\pi_Y) \qquad P(\bar{y}|\bar{A}) = 1-(1-P_1)\pi_Y$$

Clearly, $P(y|A) > P(y|\bar{A})$ if $P_1 > 0$, and by (5.4), the jeopardy function turns out to be, say,

$$g_U(y|A) = \frac{P_1 + (1-P_1)\pi_Y}{(1-P_1)\pi_Y} \qquad g_U(\bar{y}|\bar{A}) = \frac{1-(1-P_1)\pi_Y}{(1-P_1)(1-\pi_Y)}$$

Now, as before, if k_1 and k_2 be the maximal allowable values for $g_U(y|A)$ and $g_U(\bar{y}|\bar{A})$, the optimal choice of the design parameters π_Y and P_1 is seen to be given by

$$\pi_Y = \frac{k_2 - 1}{k_1 + k_2 - 2} \qquad P_1 = \frac{(k_1-1)(k_2-1)}{k_1 k_2 - 1} \tag{5.13}$$

Thus in a sense, the unrelated question model has an advantage over the Warner model since, unlike the latter, the former involves two design parameters and allows distinct upper bounds for $g(y|A)$ and $g(\bar{y}|\bar{A})$ to be attained.

It sometimes happens that $\bar{A}$ is innocuous and only A is stigmatizing. Then one can allow $k_2 \to \infty$, and by (5.12) and (5.13), the optimal choice of the design parameters for the Warner model and the unrelated question model turns out to be

$$P = \frac{k_1}{k_1 + 1} \quad \text{and} \quad P_1 = \frac{k_1 - 1}{k_1} \qquad \pi_Y = 1$$

respectively. With these choices of the design parameters, the variances of the corresponding UEs of π_A as in (1.2) and (2.2) [see also (5.8)] become

$$\text{var}(\hat{\pi}_{AW}) = \frac{\pi_A(1-\pi_A) + k_1(k_1-1)^{-2}}{n} \tag{5.14}$$

$$\text{var}(\hat{\pi}_{AU1}) = \frac{\pi_A(1-\pi_A) + (1-\pi_A)(k_1-1)^{-1}}{n} \tag{5.15}$$

Evidently, (5.15) is smaller than (5.14) so that the unrelated-question model (with known π_Y) is more efficient than the Warner model, when compared at the same level of protection of privacy, provided that $\bar{A}$ is innocuous and A and Y are independent. This is expected since the Warner model involves only one design parameter and fails to utilize the additional information that $\bar{A}$ is innocuous.

An easy way to achieve $\pi_Y = 1$ is to so design that with a probability P_1 a respondent is to divulge his or her truth about A and to be instructed to report "yes" with the complementary probability $(1 - P_1)$.

The literature on RR techniques contains a wide variety of measures for protecting privacy. We will consider a few of them here. In dealing with the unrelated-question model, Lanke (1975) calls $P(A|y)$ the "risk of suspicion." He assumes that only A is stigmatizing but $\bar{A}$ is innocent. Also, he assumes that $\pi_A < 1/2$ and takes A and Y as independent attributes. To keep the risk of suspicion under control, he suggests choosing a convenient positive number $\theta \leq 1$, demanding that $P(A|y) \leq \theta$. Hence

$$\frac{P_1\pi_A + (1-P_1)\pi_A\pi_Y}{P_1\pi_A + (1-P_1)\pi_Y} \leq \theta$$

yielding

$$P_1 \leq \frac{(\theta - \pi_A)\pi_Y}{(1-\theta)\pi_A + (\theta - \pi_A)\pi_Y} = \psi \quad \text{(say)}$$

A necessary and sufficient condition for the existence of such a P_1 is $\pi_A \leq \theta$, in which case Lanke (1975) recommends the choice $P_1 = \psi$. This ψ increases with π_Y. It follows that $\text{var}(\hat{\pi}_A)$, with $P_1 = \psi$, decreases as π_Y increases. Therefore, he approves of the use of a Y with $\pi_Y = 1$. Of course, it is difficult to settle how to fix ψ, which involves the unknown constant π_A.

Leysieffer (1975) also demonstrates the parallel result that among all unrelated-question designs with the same jeopardy level with respect to $P(A|R)$, the designs with $\pi_Y=1$ yield the least value for $\text{var}(\hat{\pi}_A)$, in case A alone but not $\bar{A}$ is stigmatizing. About the practical choice of $\pi_Y=1$, Lanke (1975) argues that since only A is stigmatizing, a respondent may not be embarrassed always to report "yes" when asked about membership in Y. This view, however, may not be tenable if one observes that since "yes" may sometimes relate to the sensitive group A, a clever respondent may, falsely but safely, always be inclined to respond "no".

In fact, the recommendation $\pi_Y=1$ may not often work well when one adopts other measures of respondent jeopardy. Greenberg *et al.* (1977) emphasize a need to control $P(A|R)$ and $P(\bar{A}|R)$ to ensure truthful cooperation of people of groups A and Ā both of which may carry a stigma. They define the hazard for A, say, H_A, as the probability that a respondent of group A is perceived as belonging to group A. Clearly,

$$H_A=P(y|A)P(A|y)+P(\bar{y}|A)P(A|\bar{y}) \tag{5.16}$$

Defining $H_{\bar{A}}$ similarly, the expected hazard is

$$\bar{H}_A=\pi_A H_A+(1-\pi_A)H_{\bar{A}} \tag{5.17}$$

Greenberg *et al.* (1977) interpret $1-H_A$ as the gain due to randomization in protecting privacy with truthful reporting from a person of group A, and $H_{\bar{A}}$ as the corresponding loss for one in Ā. They also consider two related measures of hazard from reporting "yes," namely, H_A^*, the probability that a person of group A is perceived so when reporting "yes," and $H_{\bar{A}}^*$, defined in an obviously analogous manner. Defining

$$B_A=1-H_A^* \qquad B_{\bar{A}}=1-H_{\bar{A}}^*$$

the expected overall gain due to randomization is

$$G=\pi_A B_A+(1-\pi_A)B_{\bar{A}} \tag{5.18}$$

In particular, for the unrelated-question model with known π_Y, it may be seen from (5.17) and (5.18) that

$$H_A=\frac{[P_1+(1-P_1)\pi_Y][P_1\pi_A+(1-P_1)\pi_A\pi_Y]}{P_1\pi_A+(1-P_1)\pi_Y} + \frac{[(1-P_1)(1-\pi_Y)][(1-P_1)\pi_A(1-\pi_Y)]}{P_1(1-\pi_A)+(1-P_1)(1-\pi_Y)}$$

and

$$G=\pi_A(1-\pi_Y)(1-P_1) \tag{5.19}$$

By (5.19), G decreases monotonically with P_1 and π_Y. As a guide to the choice of good designs, Greenberg *et al.* (1977) recommend maximizing $T=B_A(1-B_A)$. For fixed π_Y, the quantity T increases with P_1 so that a large P_1 is desirable, while for a fixed P_1 it decreases with π_Y so that a large π_Y is not desirable. Thus their recommendations contradict Lanke's (1975) and Leysieffer's (1975) suggestions for the choice $\pi_Y=1$.

Lanke (1976) considers yet another measure of protecting of privacy, namely, $\max[P(A|y), P(A|\bar{y})]=\mathscr{P}$, say. For the unrelated-question model with known π_Y, assuming as before that A and Y are independent, we obtain

$$P(A|y)=\frac{P_1\pi_A+(1-P_1)\pi_A\pi_Y}{P_1\pi_A+(1-P_1)\pi_Y} \tag{5.20}$$

$$P(A|\bar{y})=\frac{(1-P_1)\pi_A(1-\pi_Y)}{P_1(1-\pi_A)+(1-P_1)(1-\pi_Y)}$$

and it may be checked that $P(A|y)>P(A|\bar{y})$. The same is easily seen to be true for the Warner model if $P>1/2$, which may be assumed without loss of generality. Lanke (1976) therefore takes $\mathscr{P}=P(A|y)$. Denoting $\mathscr{P}$ for the Warner model and the unrelated-question model by $\mathscr{P}_W$ and $\mathscr{P}_U$, respectively, these two models will be considered equivalent from the point of view of protection of privacy if $\mathscr{P}_W=\mathscr{P}_U$. Since

$$\mathscr{P}_W=\frac{P\pi_A}{P\pi_A+(1-P)(1-\pi_A)}$$

$$\mathscr{P}_U=\frac{P_1\pi_A+(1-P_1)\pi_A\pi_Y}{P_1\pi_A+(1-P_1)\pi_Y}$$

it follows that for every P_1 and π_Y, there exists a unique value of P, namely,

$$P=\frac{1}{2}+\frac{P_1}{2P_1+4(1-P_1)\pi_Y} \tag{5.21}$$

such that $\mathscr{P}_W=\mathscr{P}_U$ for all π_A (i.e., the two models are equally protective). For given P_1 and π_Y, suppose that P is chosen according to (5.21). Then considering the variances of the UEs of π_A for the two models as in (1.3) and (2.3), and applying (5.21), it can be shown that

$$\text{var}(\hat{\pi}_{AW})-\text{var}(\hat{\pi}_{AU1})=\frac{(1-P_1)(2\pi_Y-1)}{1-\lambda_U}\text{var}(\hat{\pi}_{AU1})$$

where $\lambda_U=P_1\pi_A+(1-P_1)\pi_Y$, and consequently $\text{var}(\hat{\pi}_{AW})\gtrless\text{var}(\hat{\pi}_{AU1})$ according as $\pi_Y\gtrless 1/2$. Similar conclusions can be drawn if we follow the approach of Flinger *et al.* (1977) to take

$$J_1 = \frac{1-\max\{P(A|y), P(A|\bar{y})\}}{1-\pi_A} = \frac{1-\mathscr{P}}{1-\pi_A}$$

as a measure of "primary protection" and compare the two models with respect to efficiency at the same level of J_1. Lanke (1976) also makes some comparisons between the Warner model and those considered by Moors (1971) and Folsom *et al.* (1973), taking both efficiency and protection of privacy into account.

Apart from suggesting the measure J_1 stated above, Flinger *et al.* (1977) consider another approach having a strong similarity with that due to Lanke (1976). Let the expressions for $P(A|y)$ under the Warner and unrelated-question models be denoted by $P_W(A|y)$ and $P_U(A|y)$, respectively. Similarly, define $P_W(A|\bar{y})$ and $P_U(A|\bar{y})$. According to Flinger *et al.* (1977), the two models afford equal treatment, in terms of protection of confidentiality, if (i) $P_W(A|y)=P_U(A|y)$ and (2) $P_W(A|\bar{y})=P_U(A|\bar{y})$. As usual, assuming the independence of A and Y, observe that

$$P_W(A|y) = \frac{P\pi_A}{P\pi_A+(1-P)(1-\pi_A)}$$

$$P_W(A|\bar{y}) = \frac{(1-P)\pi_A}{(1-P)\pi_A+P(1-\pi_A)}$$

and that $P_U(A|y)$ and $P_U(A|\bar{y})$ are as given by (5.20). Hence it can be shown, after some algebra, that the two models are equivalent [i.e., (1) and (2) hold] in terms of protection to the respondent if $P_1=2P-1$ and $\pi_Y=1/2$. With this choice of P_1 and π_Y, both models have the same probability of a "yes" response, so that the sample proportion of "yes" responses for the two models have a common distribution. As a consequence, the corresponding usual UEs for π_A based on the two models become equally efficient. Thus under this criterion of Flinger *et al.* (1977), equivalence in terms of protection of privacy implies that in terms of efficiency.

In concluding this section, we may mention that Warner (1976) reports certain results similar to those by Leysieffer and Warner (1976) already presented in this section. He has a few additional results of interest as well. We omit them here, only drawing the interested reader's attention to them.

5.3. GENERAL RR MODELS WITH DICHOTOMOUS POPULATION

Anderson (1976) extends Lanke's (1975, 1976) ideas in the following manner. He supposes that individuals of groups A and Ā respond according to known probability density (or mass) functions f_0 and f_1, respectively. He calls f_0 and f_1 the response densities. For an RR device, a respondent gives a response according to a probability density (or mass) function which is a mixture of f_0 and f_1. For a dichotomous population divided into groups A and $\bar{A}$, let a typical response from a subject be R and the totality of responses be $\mathscr{R}$. Then the responses appear in accordance with the mixture probability law,

$$f(R) = \pi_A f_0(R) + (1 - \pi_A) f_1(R) \qquad R \in \mathscr{R} \tag{5.22}$$

Analogously to Lanke's (1975, 1976) concept of risks of suspicion, Anderson (1976) considers

$$P(A|R) = \frac{\pi_A f_0(R)}{f(R)} \qquad P(\bar{A}|R) = 1 - P(A|R) \qquad R \in \mathscr{R},$$

and observes that for a greater protection of privacy, one should aim at smaller values of $|P(A|R) - \pi_A|$ and $|P(\bar{A}|R) - (1 - \pi_A)|$. Since

$$E[P(A|R)] = \pi_A E\left[\frac{f_0(R)}{f(R)}\right] = \pi_A \int f_0(R)\mathrm{d}R = \pi_A$$

it follows that the less the magnitude of $\mathrm{var}(P(A|R))$, the greater the protection of privacy.

It may be seen that

$$\mathrm{var}(P(A|R)) = \pi_A^2 E\left[\frac{f_0(R)}{f(R)} - 1\right]^2$$

Hence by (5.22), assuming $f(R)$ to satisfy the usual regularity conditions, Anderson (1976) works out the Fisher information under this model as

$$I = E\left[\frac{\mathrm{d}}{\mathrm{d}\pi_A} \log f(R)\right]^2 = E\left[\frac{f_0(R) - f_1(R)}{f(R)}\right]^2 = \frac{\mathrm{var}(P(A|R))}{[\pi_A(1-\pi_A)]^2} \tag{5.23}$$

For any UE, say $\tilde{\pi}_A$, for π_A the Rao-Cramér inequality gives

$$\mathrm{var}(\tilde{\pi}_A) \geq \frac{1}{nI} = \frac{[\pi_A(1-\pi_A)]^2}{n\ \mathrm{var}(P(A|R))}$$

where n is the sample size.

It is important to observe that the less the variability in the risk of suspicion, the greater the protection of confidentiality. But from (5.23) it follows that with decreasing variability in the risk of suspicion there is diminishing information and increasing variance (i.e., decreasing efficiency for unbiased estimators for π_A).

For the Warner model one may check that

$$\text{var}(P(A|R)) = \frac{[\pi_A(1-\pi_A)]^2(2P-1)^2}{\lambda(1-\lambda)}$$

where $\lambda = (1-P) + (2P-1)\pi_A$. Hence, by (5.23),

$$I = \frac{(2P-1)^2}{\lambda(1-\lambda)}$$

On the other hand, by (1.3),

$$\text{var}(\hat{\pi}_{AW}) = \frac{\lambda(1-\lambda)}{n(2P-1)^2}$$

Therefore, $\text{var}(\hat{\pi}_{AW}) = 1/nI$, and the Warner estimator attains the Rao-Cramér lower bound. In the unrelated-question model with a known π_Y, the UE for π_A, namely $\hat{\pi}_{AU1}$[see (2.2)] also attains the Rao-Cramér lower bound, as one may check following Anderson's (1976) arguments based on Boes' (1966) result on mixture distributions.

5.4. POLYCHOTOMOUS MODELS

In this section we consider some of the ideas discussed in the preceding sections as extended by Anderson (1975a) and Loynes (1976) to the multiproportions case. Consider a population classified into t mutually exclusive and exhaustive classes $A_1, \ldots, A_t$, according to some sensitive character, with corresponding unknown proportions $\pi_1, \ldots, \pi_t$ $(\Sigma_t^t \pi_j = 1)$. With an SRSWR of size n, let $\underline{Z}$ be the set of all possible randomized responses and z in $\underline{Z}$ a typical response. For $j = 1, \ldots, t$, let $P(z|A_j)$ denote the conditional probability that an individual in group or class A_j makes the response z. These quantities $P(z|A_j)$ are, of course, at the investigator's disposal.

For example, with the Bourke and Dalenius (1976) scheme (see Section 3.3), $\underline{Z} = \{1, 2, \ldots, t\}$, and for $z \in \underline{Z}$; $j = 1, 2, \ldots, t$,

$$\begin{aligned} P(z|A_j) &= p + p_j \qquad \text{for } z = j \\ &= p_z \qquad \text{for } z \neq j \end{aligned}$$

Loynes (1976) extends the notion of the jeopardy function of Leysieffer and Warner (1976) to polychotomous populations. Suppose the design parameters are such that if $P(z|A_j)>0$ for some $j(=1, \ldots, t-1)$; then $P(z|A_t)>0$. Then the quantities

$$g_j(z)=\frac{P(z|A_j)}{P(z|A_t)} \qquad (j=1, \ldots, t) \tag{5.24}$$

are well defined for every j and every z such that $P(z|A_j)>0$. Trivially, $g_t(z)=1$ and it is easy to recognize (5.24) as a generalization of the jeopardy function g defined in (5.4) considered in the context of dichotomous populations. Since by Bayes' theorem,

$$P(A_j|z)=\frac{\pi_j P(z|A_j)}{\sum_{j=1}^{t} \pi_j P(z|A_j)}$$

it follows from (5.24) that for $j=1, \ldots, t$; $z \in \underline{Z}$,

$$P(A_j|z)=\frac{\pi_j g_j(z)}{\pi_t+\sum_{j=1}^{t-1} \pi_j g_j(z)}$$

The quantities $P(z|A_j)$ define a design. For a particular RR design d, one may write $P_d(z|A_j)$ for $P(z|A_j)$. Similarly, if the quantities $P(A_j|z)$ for a design d are denoted by $P_d(A_j|z)$, the risk $r(d)$ of the design may be defined as the vector $(r_1, \ldots, r_{t-1})'$ or $(r_1, \ldots, r_t)'$, as appropriate, where

$$r_j=\max_{z \in \underline{Z}} \quad P_d(A_j|z)$$

A design d_0, say, carries no less risk of suspicion than another design d_1 if

$$\underline{r}(d_0) \geq \underline{r}(d_1)$$

in the sense that

$$r_j(d_0) \geq r_j(d_1) \qquad \text{for each } j=1, \ldots, t$$

For the sake of notational simplicity, hereafter we shall suppress the subscript d in $P_d(z|A_j)$. Now for any $z \in \underline{Z}$, the overall probability of the response z is $\Sigma_{j=1}^{t} \pi_j P(z|A_j)$.

So, with an SRSWR of size n, denoting the response from the uth respondent by $z_u(u=1, \ldots, n)$, the likelihood function is seen to be given by

$$\log L=\sum_{u=1}^{n} \log\left[\sum_{j=1}^{t} \pi_j P(z_u|A_j)\right]$$

Recalling that $\Sigma_{j=1}^{t} \pi_j=1$, it may be seen that for $j, j'=1, \ldots, t-1$,

$$\frac{\partial^2 \log L}{\partial \pi_j \, \partial \pi_{j'}} = \sum_{u=1}^{n} \frac{[P(z_u|A_j) - P(z_u|A_t)][P(z_u|A_{j'}) - P(z_u|A_t)]}{\left[\sum_{j=1}^{t} \pi_j P(z_u|A_j)\right]^2}$$

Applying (5.24), one gets the $(t-1) \times (t-1)$ information matrix $I = (I_{jj'})$ as

$$I_{jj'} = E\left(-\frac{\partial^2 \log L}{\partial \pi_j \, \partial \pi_{j'}}\right)$$

$$= n \sum_{z \in Z} \frac{P(z|A_t)[g_j(z) - 1][g_{j'}(z) - 1]}{\pi_t + \sum_{j=1}^{t-1} \pi_j g_j(z)}$$

The asymptotic dispersion matrix of the maximum likelihood estimators of $\pi_1, \ldots, \pi_{t-1}$ is I^{-1} (the inverse of matrix I). So, denoting the information matrix of a design d by $I(d)$, a design d_0 is at least as good as another design d_1 from the point of view of efficiency if $I(d_0) - I(d_1)$ is nonnegative definite.

A design d_0 will be called admissible if there exists no other design that dominates d_0 in terms of both efficiency and risks of suspicion, that is, if there does not exist a design d for which (1) $\underline{r}(d) \leq \underline{r}(d_0)$, and (2) $I(d) - I(d_0)$ is nonnegative definite, with either $\underline{r}(d) \neq \underline{r}(d_0)$ or $I(d) \neq I(d_0)$, or both.

Observe that each possible response z may be represented by a point

$$\underline{g}(z) = (g_1(z), \ldots, g_{t-1}(z))'$$

and thus any given design has a representation as a set of points in the $(t-1)$-dimensional space. Suppose that the set of points constituting a given design has an internal point that is, a point which can be written as a convex combination of some other points in the design. Loynes (1976) shows that such a design is inadmissible by demonstrating that it is possible to construct another design which is as good as the former in terms of risk but better in terms of efficiency.

As an alternative approach, Anderson (1975a) extends (5.11) to the multiproportions case and recommends that for protection of confidentiality the investigator should prescribe a $(t-1)$-dimensional set in which the vector $(P(A_1|z), \ldots, P(A_{t-1}|z))'$ should lie for each $z \in \underline{Z}$ for some guessed values of $\boldsymbol{\pi} = (\pi_1, \ldots, \pi_t)'$. This set, Q_π, is called the risk restriction set. A design d satisfying the foregoing restriction will be called a Q_π plan. Anderson (1975a) defines a Q_π plan d_0 admissible if there does not exist another Q_π plan d such that $I(d) - I(d_0)$ is nonnegative definite with $I(d) \neq I(d_0)$.

If Q_π is a convex simplex, it is shown by Anderson (1975a) that the vector $(P(A_1|z), \ldots, P(A_{t-1}|z))'$ for an admissible Q_π plan takes values on the vertices of Q_π with probability 1. Further, the optimal Q_π plan for estimating a specified linear combination of $\pi_1, \ldots, \pi_t$ has at most $2(t-1)$ possible responses. Anderson (1975a) also treats proportions depending on a parameter θ and shows that when θ is one-dimensional, determination of the optimal Q_π plan leads to a linear programming problem.

5.5. ADDITIONAL GENERALITIES

Anderson (1976, 1977) has further general considerations for planning RR techniques keeping in mind the effects of maintenance of confidentiality on the information content of realized survey data and on the resulting efficiency of the estimation procedure. Let X be a variable of interest and our problem be to estimate a parameter θ involved in the distribution of X. Suppose that X is not observable directly but that through an RR device one generates observations on a response variable Y such that the conditional distribution of Y given X does not involve θ.

In order to examine the relative efficiency of an RR survey method compared to a direct response (DR) method, Anderson proceeds as follows. Assume that the distributions of X, Y satisfy standard regularity conditions. Let $I(Y)$ be Fisher's information per unit observation on Y and $I(X)$ be that per observation if X was observable. Since the distribution of Y given X does not involve θ, one obtains

$$I(X) \geq I(Y)$$

Consequently, the loss of information because of the adoption of an RR method rather than a DR method is given by, say,

$$L = 1 - \frac{I(Y)}{I(X)} \tag{5.25}$$

which is nonnegative.

Turning to the problem of protection of privacy, Anderson (1976) considers the notion of revealing density which is the conditional density (or mass) function of X given a particular response which is an observation on Y. If this conditional distribution is closely concentrated around some particular value of X, the protection of privacy is very little. Therefore, $\text{var}(X \mid Y = y)$ may be taken as a measure of protection of privacy for a given y. Also, $E(\text{var}(X \mid Y))$ may be interpreted as an overall measure of protection of privacy. Various other reasonable measures of protection of privacy may be thought of in this context. For example $\min_y \operatorname{var}(X \mid Y=y)$ may be considered as one such measure.

As usual, it is seen that maintenance of confidentiality and efficient estimation are in conflict. The following example serves as an illustration. Let X have the normal distribution with unknown mean θ and variance unity. An RR survey is so designed that a respondent is asked to take an observation Z from a normal distribution with zero mean and a preassigned variance σ^2 and then report the value $Y=X+Z$. Note that X and Z are independent and Y has the normal distribution with mean θ and variance $1+\sigma^2$. It is easily seen that $I(X)=1$, $I(Y)=1+\sigma^2$, and hence, by (5.25),

$$L=\frac{\sigma^2}{1+\sigma^2}$$

Also, one may check that the conditional variance of X given $Y=y$ is $\sigma^2/(1+\sigma^2)$. Hence

$$E[\operatorname{var}(X \mid Y)]=\frac{\sigma^2}{1+\sigma^2}$$

so that the loss of information due to adoption of RR equals the chosen measure of protection of privacy. In fact, a large value of σ^2 instills a high sense of protection of privacy but at the same time, leads to an increased loss of information.

Anderson (1975b) also discusses the problem of testing two simple hypotheses by means of RR. For a fixed level of significance, the loss of power due to RR is estimated by means of Kullback-Leibler's information. As in the case of the estimation problem, the loss of information equals the expected conditional Kullback-Leibler information from X given Y.

As we do not have access to the work in the areas relevant to this chapter reported in Dutch by Albers (1982), we cannot discuss it. Similarly, a review on RR by Verdooren (1976), also in Dutch, is beyond our coverage.

REFERENCES

Albers, W. (1982). Simple randomized response procedures with bounded respondent risk for quantitative data. *Kwantitatieve Methoden* **8**, 35–46.

Anderson, H. (1975a). Efficiency versus protection in the RR designs for estimating proportions. *Tech. Rep.* 9, University of Lund, Sweden.

Anderson, H. (1975b). Efficiency versus protection in a general RR model. *Tech. Rep.* 10, University of Lund, Sweden.

Anderson, H. (1975c). Efficiency versus protection in RR designs. Mimeo notes, University of Lund, Sweden.

Anderson, H. (1976). Estimation of a proportion through RR. *Internat. Statist. Rev.* **44**, 213–217.

Anderson, H. (1977). Efficiency versus protection in a general RR model. *Scand. J. Statist.* **4**, 11–19.

Boes, D. C. (1966). On the estimation of mixing distribution. *Ann. Math. Statist.* **37**, 177–188.

Flinger, M. A., Policello, G. E., and Singh, J. (1977). A comparison of two RR survey methods with consideration for the level of respondent protection. *Comm. Statist.—Theory Methods* **6**, 1511–1526.

Folsom, R. E., Greenberg, B. G., Horvitz, D. G., and Abernathy, J. R. (1973). The two alternate questions RR model for human surveys. *J. Amer. Statist. Assoc.* **68**, 525–530.

Greenberg, B. G., Kubler, R. R., Abernathy, J. R., and Horvitz, D. G. (1977). Respondent hazards in the unrelated question RR model. *J. Statist. Plann. Inference* **1**, 53–60.

Lanke, J. (1975). On the choice of unrelated question in Simmons' version of RR. *J. Amer. Statist. Assoc.* **70**, 80–83.

Lanke, J. (1976). On the degree of protection in randomized interviews. *Internat. Statist. Rev.* **44**, 197–203.

Lehmann, E. H. (1959). *Testing Statistical Hypotheses.* John Wiley, New York.

Leysieffer, R. W. (1975). Respondent jeopardy in RR procedures. *Tech. Rep. M*338, Department of Statistics, Florida State University, Tallahassee.

Leysieffer, R. W., and Warner, S. L. (1976). Respondent jeopardy and optimal designs in RR models. *J. Amer. Statist. Assoc.* **71**, 649–656.

Loynes, R. M. (1976). Asymptotically optimal RR procedures. *J. Amer. Statist. Assoc.* **71**, 924–928.

Moors, J. J. A. (1971). Optimization of the unrelated question RR model. *J. Amer. Statist. Assoc.* **66**, 627–629.

Verdooren, L. R. (1976). Loten bij delicate vragen; een overzicht van "randomized response"—technicken (in Dutch). *Statist. Neerlandica* **30**, 7–24.

Warner, S. L. (1976). Optimal RR models. *Internat. Statist. Rev.* **44**, 205–212.

6

Miscellaneous Topics on RR Techniques

6.1. A BAYESIAN APPROACH

Winkler and Franklin (1979) describe a Bayesian approach in the context of Warner's (1965) RR model. For a dichotomous population with π_A as the proportion of persons bearing a sensitive character A, let, following Warner (1965), an RR be employed with P as the probability that a subject answers "yes" or "no" about bearing A and with probability $(1-P)$ about its complement. Let λ be the probability of a "yes" response. For an SRSWR of size n the likelihood is

$$L = L(\lambda) = \binom{n}{n_1} \lambda^{n_1} (1-\lambda)^{n-n_1} \tag{6.1}$$

where n_1 is the frequency of "yes" replies and

$$\lambda = P\pi_A + (1-P)(1-\pi_A)$$

For simplicity, let us assume that $P > 1/2$. Since $\pi_A \in [0, 1]$, one has $\lambda \in [1-P, P]$. In estimating π_A, Winkler and Franklin (1979) take a natural conjugate beta form of a prior for π_A with a density

$$f_\beta(\pi_A | \alpha', \beta') = \frac{1}{B(\alpha', \beta')} \pi_A^{\alpha'-1} (1-\pi_A)^{\beta'-1} \tag{6.2}$$

Combining (6.1) with (6.2), one has the posterior for π_A as

$$f(\pi_A | n_1, n) \propto \pi_A^{\alpha'-1} (1-\pi_A)^{\beta'-1} [(1-P) + (2P-1)\pi_A]^{n_1} \times [P - (2P-1)\pi_A]^{n-n_1} \tag{6.3}$$

Expanding the terms within brackets, one may simplify this posterior to the form

$$f(\pi_A|n_1, n)=\sum_{t=0}^{n} W_t f_\beta(\pi_A|\alpha'+t, \beta'+n-t) \qquad 0\leq\pi_A\leq 1 \qquad (6.4)$$

where $W_t = W_t^*/(\Sigma_{s=0}^{n} W_s^*)$ and

$$W_t^* = \binom{n}{t}\frac{B(\alpha'+t, \beta'+n-t)}{B(\alpha', \beta')}\sum_{j=0}^{\min(n_1, t)}\binom{t}{j}\binom{n-t}{n_1-j}P^{n-t-n_1+2j} \times (1-P)^{t+n_1-2j}$$

Thus the posterior is a mixture of beta distributions. Each term in the sum (6.4) is the product of the posterior probability that exactly t of the n respondents actually belong to group A and the posterior distribution of π_A given that exactly t respondents actually bear A. Because of the RR plan t is never observed, and this uncertainty about t causes the posterior distribution of π_A to be a mixture of $(n+1)$ beta distributions instead of a single beta distribution.

The Bayes estimator of π_A, which is the mean of the posterior distribution (6.4), turns out as

$$\hat{\pi}_{AB}=\sum_{t=0}^{n}\frac{W_t(\alpha'+t)}{\alpha'+\beta'+n}$$

The weights W_t are, however, somewhat involved in nature. So certain approximations have been given by Winkler and Franklin (1979) for simplifications. We omit them here, drawing a curious reader's attention to the original reference cited.

Since the moment estimator $\hat{\lambda}=n_1/n$ of λ may yield values outside the parametric space $[1-P, P]$ for λ with a range $(2P-1)$ and hence may lead to Warner's estimator $\hat{\pi}_{AW}$ assuming a value outside [0,1] for certain choices of P, Mukhopadhyay and Halder (1980) suggested (1) calling $I(P)=2P-1$ an index of sensitivity of an RR experiment, and (2) seeking a Bayes estimate for λ and hence for π_A on postulating a beta prior for λ over $[1-P, P]$, assuming that $P>1/2$. As an alternative they also recommended, in case no prior information is available, a uniform prior for λ with a density

$$g(\lambda)=\frac{1}{2P-1} \qquad \text{if } \lambda\in[1-P, P]$$
$$=0 \qquad \text{otherwise}$$

These authors have evaluated the variances of the resulting Bayes estimates $\hat{\lambda}_b$ for λ and $\hat{\pi}_b$ for π_A and tabulated the latter for various n, n_1, and P.

Gunel (1985) postulated a beta prior for π_A in the Warner model, considered a square error loss, obtained Bayes' estimators, and compared Bayes' risks under the Warner RR model and DR models allowing nonresponses with the latter. With other models as well, it is worth obtaining Bayes' estimates and studying them under RR models.

6.2. MORE LYING MODELS

Abul-Ela *et al.* (1967) allowed false RR leading to biased estimators and examined the extent of biases and the magnitudes of the resulting mean square errors of estimators in multiproportion cases. They gave details with a trichotomy. Bourke and Dalenius (1974) consider estimating π_A on applying Warner's (1965) original RR scheme under the latter model but suppose that truthful responses may come not from everybody bearing A but only from a proportion $H(P)$, say, of them, namely, from among the honest ones. They call $H(P)$ the "honesty function." Some details in this regard were given in Exercise 1.3.

Mukhopadhyay (1980) considers the following lying model. He illustrates a case when a population is divided into four categories $i = 1, 2, 3, 4$ in respect to a certain qualitative character. Let the categories bear social stigma in a decreasing order from 1 to 4 and the category 4 be innocuous, but for simplicity let 2 and 3 be equally embarrassing. Let T_{ij} be the probability that a person of category i, in a direct survey in this context, announces himself or herself as one of category j ($i, j = 1, 2, 3, 4$). It is intuitively clear that one should stipulate that

$$T_{21} = T_{23} = T_{31} = T_{32} = T_{41} = T_{42} = T_{43} = 0$$

$$T_{22} + T_{24} = 1 \qquad T_{33} + T_{34} = 1 \qquad T_{44} = 1 \qquad \sum_{j=1}^{4} T_{1j} = 1$$

Letting $r_i (i = 1, \ldots, 4)$ denote the sample proportion, in an SRSWR of size n, of responses divulging membership of category i, Mukhopadhyay (1980) gives moment estimators (wrongly claimed by him as maximum likelihood estimators) for π_1 and π_2, the proportions in categories 1 and 2. He finds the estimators biased and examines their variances and mean square errors. He demonstrates with numerical examples that the more a society deviates from truthfulness, the greater the efficiency of an estimator based on an RR survey over that on an open survey. He extends this model to cover a case where an RR survey is conducted and yet the respondents may lie. Let the probability T_{ij} in the previous DR now be denoted as T'_{ij} in an RR counterpart. Then one would demand

$$T'_{21}=T'_{23}=T'_{31}=T'_{32}=T'_{41}=T'_{42}=T'_{43}=0$$

$$T'_{22}+T'_{23}=1 \qquad T'_{33}+T'_{34}=1 \qquad T'_{44}=1 \qquad \sum_{j=1}^{4} T'_{1j}=1$$

In this case also he derives moment estimators, finds them biased, and obtains the MSEs. He considers it reasonable to assume that

$$T'_{ii}>T_{ii} \qquad (i=1,\ldots,4)$$

and empirically studies the relative efficiencies of estimators based on direct response (DR) and RR, in this case allowing lies in either type of survey technique. He illustrates situations to show that when lying is permitted it may be so modeled as to suggest that an RR technique may yet prove more useful than a DR. More general cases covering any finite number of categories have been investigated by Mukerjee (1981).

6.3. RANDOMIZED RESPONSE SURVEYS ALLOWING OPTIONS FOR DIRECT RESPONSES

Both intuition and experience suggest that certain items of inquiry apprehended by an investigator to be socially stigmatizing in general may not be so felt by certain sampled subjects. For most of the currently available techniques of applying RR schemes, one has to insist on getting an RR even from a volunteer ready to give a DR and forgo a piece of valuable information only for the sake of deriving an unbiased estimator with certain analytical properties. Recently, Chaudhuri and Mukerjee (1985) therefore suggest a plan to get optional randomized responses (ORR) rather than compulsory ones (CRR, say), allowing each respondent to opt either for a DR or an RR. The resulting gain in efficiency because of this flexibility emerges through a version of Rao-Blackwellization. This is achievable with each RR technique, if suitably modified into an ORR technique. But these authors have illustrated only certain specific cases where it is shown explicitly to work. The method should be successful in practice if a character under study, A, say, is not associated with a respondent's choice for DR or RR, so that just an option for RR does not create a respondent jeopardy.

The authors claim that their method yields an ORR equally protective of confidentiality as a CRR but with enhanced efficiency. They illustrate the classical case of a multinomial sampling as follows. Let a population consist of individuals of t distinct categories A_j with proportions π_j $(0<\pi_j<1)$, $j=1, \ldots, t, \Sigma\pi_j=1$, according to a sensitive character A. The problem is to estimate the $u \times 1$ vector

$$\theta = G\pi$$

where $\pi=(\pi_1, \ldots, \pi_t)'$, u is any positive integer, and G is any known $u \times t$ matrix. Let $m(\geq 1)$ independent SRSWRs of sizes n_i $(i=1, \ldots, m, \Sigma n_i=n)$ be taken. Following some standard technique for the ith sample, let, $r_i(\geq 1)$ distinct RRs be available according to some known $r_i \times t$ design matrix P_i with nonnegative entries, $i=1, \ldots, m$. We will denote by I_r the $r \times r$ identity matrix, $\underline{1}_r$ the $r \times 1$ column vector with 1 as each entry, and α^δ, a diagonal matrix with diagonal entries as elements of a vector α.

$$\underline{1}'_{r_i} P_i = \underline{1}'_{t'} \qquad (i=1, \ldots, m)$$

Let $\mu_i = P_i\pi$ be the $r_i \times 1$ vector of probabilities of the different RRs with $\hat{\mu}_i$ as the corresponding vector of observed proportions, $i=1, \ldots, m$. Let

$$P=(P'_1, \ldots, P'_t)'$$

be a matrix of order $(\Sigma r_i) \times t$.

A necessary and sufficient condition for the existence of a UE for θ in terms of $\hat{\mu}_i$, $i=1, \ldots, m$, is the existence of a matrix B such that

$$G = BP$$

Writing $B=(B_1, \ldots, B_m)$, where B_i is a $u \times r_i$ matrix, a UE of θ is

$$\hat{\theta}_c = \sum_1^m B_i \hat{\mu}_i$$

Its dispersion matrix is

$$V_1 = \sum_{1=1}^m \frac{B_i(\mu_i^\delta - \mu_i \mu'_i) B'_i}{n_i}$$

Suppose now that DRs as well as RRs are allowed. Then let

$$\pi = \pi^{(1)} + \pi^{(2)}$$

where $\pi^{(1)}$ $(\pi^{(2)})$ is the vector of unknown proportions of categories A_j's and opting for DR (RR). For each respondent in the ith sample there are $(t+r_i)$ possible responses—either one of the t DRs stating his or her true category or one of the r_i RRs according to the design matrix P_i, the corresponding $(t+r_i) \times 1$ vector of probabilities being $(\pi^{(1)'}, \mu_i^{(2)'})$, where

$$\mu_i^{(2)} = P_i \pi^{(2)} \qquad (i=1, \ldots, m)$$

Let $\hat{\pi}_i^{(1)}$ and $\hat{\mu}_i^{(2)}$ denote vectors of observed proportions corresponding to $\pi_i^{(1)}$ and $\mu_i^{(2)}$. Consider

$$\hat{\theta}^{(0)} = \sum_{i=1}^{m} B_i(P_i\hat{\pi}_i^{(1)} + \hat{\mu}_i^{(2)})$$

which is the conditional expectation of $\hat{\theta}_c$ over randomization only for the sampled units opting for DR. This $\hat{\theta}^{(0)}$ is defined from $\hat{\theta}_c$ through Rao-Blackwellization. Naturally, writing V_2 for the dispersion matrix of $\hat{\theta}^{(0)}$, one may conclude that $V_1 - V_2$ is nonnegative definite, implying the superiority of $\hat{\theta}^{(0)}$ over $\hat{\theta}_c$. One may work out

$$V_2 = \textstyle\sum_{i=1}^{m} B_i(P_i\pi^{(1)\delta}P_i' + \mu_i^{(2)\delta} - P_i\pi\pi'P_i')B_i'/n_i$$

Writing $\mu_i^{(1)} = \mu_i - \mu_i^{(2)} = P_i\pi^{(1)}$, we have

$$V_1 - V_2 = \sum_{i=1}^{m} \frac{B_i(\mu_i^{(1)\delta} - P_i\pi^{(1)\delta}P_i')B_i'}{n_i}$$

which may be directly verified to be nonnegative definite by Lemma 3.1.

6.4. SOME ALLIED METHODS FOR SENSITIVE CHARACTERS

Although RR techniques are most widely applied in deriving data on stigmatizing characters, recently some other methods have received some attention. Let us review some of these procedures.

Raghavarao and Federer (1979) refer to an earlier black-box device as an alternative to the RR technique in getting data from respondents without disclosing their identity and hence preserving secrecy. This device of collecting filled-in questionnaires from undisclosed respondents in a box and randomizing through shuffling may yield more efficient results, because of possibly higher participation, than through the RR technique, provided that anonymity is convincingly preserved. That the RR method rather than the black-box device survives is, however, an obvious hint that the proviso hardly holds.

Raghavarao and Federer (1979) also demonstrate the use of block designs in collecting data on incriminating matters. Suppose that there are v questions some of which are sensitive. Each respondent reports only a total score for a set of k of the v questions ($k \leq v$). There are b different sets of questions according to some known experimental block design. A block (i.e., a set of questions) is randomly assigned to a respondent with the stipulation that all blocks have an equal or approximately equal number of respondents. The block totals for the responses form a basis for obtaining estimates of the parameters of interest. With properly constructed scores of responses, the privacy of the interviewee may be maintained. Raghavarao and Federer (1979) present some illustrations of the foregoing technique, one of which will be described below.

Suppose that among the v questions exactly one (say, the first one) is sensitive. With a total sample of size vn, it may be decided that the sensitive question will be asked of each respondent and each of the remaining $(v-1)$ questions will be asked of $2n$ individuals. Let there be $b=v$ blocks of questions with $(v-1)$ blocks containing exactly $k=2$ questions each and the remaining block containing all the v questions. Each block is assigned randomly to n respondents. Let Y_j denote the mean response (i.e., the mean of the total responses, taken over the n respondents) corresponding to the jth block ($j=1, \ldots, v$). The parameters to be estimated by the v questions will be denoted by $\theta_1, \ldots, \theta_v$. The setup is then as exhibited in Table 6.1.

If the Y_j's have a common variance, say, σ^2/n, then the least-squares estimators for $\theta_1, \ldots, \theta_v$ are as follows:

$$\hat{\theta}_1 = \frac{\sum_{j=1}^{v-1} Y_j - Y_v}{v-2}$$

$$\hat{\theta}_i = \frac{(v-3)Y_{i-1} + Y_v - \sum_{j=1, \neq i-1}^{v-1} Y_j}{(v-2)} \qquad (i=2, \ldots, v)$$

Also,

$$\text{var}(\hat{\theta}_1) = \frac{\sigma^2 v}{n(v-2)^2}$$

$$\text{var}(\hat{\theta}_i) = \frac{\sigma^2(v^2-5v+8)}{n(v-2)^2} \qquad (i=2, \ldots, v)$$

TABLE 6.1

Block	Question in the set	Block size	Response	Expected value of response
1	1, 2	2	Y_1	$\theta_1+\theta_2$
2	1, 3	2	Y_2	$\theta_1+\theta_3$
3	1, 4	2	Y_3	$\theta_1+\theta_4$
$\vdots$		$\vdots$	$\vdots$	$\vdots$
$v-1$	1, v	2	Y_{v-1}	$\theta_1+\theta_{v-1}$
v	1, 2, ..., v	v	Y_v	$\theta_1+\cdots+\theta_v$

The ideas above have been generalized by Raghavarao and Federer (1979), who describe a general "supplemented block total response" method as follows. Thus if N^* is the incidence matrix of a design D involving $v-u$ treatments in b^* blocks of sizes $k_1^*, \ldots, k_{b^*}^*$, the supplemented design D^* with u supplemented treatments has the incidence matrix

$$M = \begin{bmatrix} N^* \\ E_{ub^*} \end{bmatrix}$$

where E_{uw} is a $u \times w$ matrix with all elements unity. The u supplemented treatments in D^* may be identified with the u sensitive questions.

The parameters corresponding to not more than one sensitive question can be estimated by D^* as above. This indicates that u cannot exceed unity. Also, the parameter corresponding to even one sensitive question cannot be estimated by D^* if k_j^*'s are all equal.

In view of the foregoing observation, one may, for the sake of simplicity, make use of equireplicate pairwise balanced designs or symmetric unequal block arrangements (Raghavarao, 1971) with b^* blocks and $4t-2$ treatments, such that each treatment is replicated r^* times and each pair of distinct treatments occurs together in exactly λ^* blocks, to construct supplemented designs. If the sensitive question corresponds to the supplemented treatment $4t-1$, it can be shown that the corresponding parameter is estimable with a variance

$$\sigma^2\left[b^* - \frac{(4t-2)r^{*2}}{r^* + \lambda^*(4t-3)}\right]$$

Raghavarao and Federer (1979) also discuss the general theory of block total response technique with respect to balanced incomplete block designs.

REFERENCES

Abul-Ela, Abdel-Latif, A., Greenberg, B. G., and Horvitz, D. G. (1967). A multi-proportions RR model. *J. Amer. Statist. Assoc.* **62**, 990–1008.

Chaudhuri, A., and Mukerjee, R. (1985). Optionally randomized response techniques. *Bull. Calcutta Statist. Assoc.* **34**, 225–229.

Gunel, E. (1985). A Bayesian comparison of randomized and voluntary response sampling models. *Comm. Statist.—Theory Methods* **14**, 2411–2435.

Mukerjee, R. (1981). Inference on confidential characters from survey data. *Bull. Calcutta Statist. Assoc.* **30**, 77–88.

Mukhopadhyay, P. (1980). On estimation of some confidential characters from survey data. *Bull. Calcutta Statist. Assoc.* **29**, 133–141.

Mukhopadhyay, P., and Halder, A. K. (1980). Bayesian tables for Warner's RR probabilities. *Tech. Rep. ASC*/80/2, Indian Statistical Institute, Calcutta.

Raghavarao, D. (1971). *Constructions and Combinatorial Problems in Design of Experiments*. John Wiley, New York.

Raghavarao, D., and Federer, W. T. (1979). Block total response as an alternative to RR method in surveys. *J. Roy. Statist. Soc. Ser. B* **41**, 40–45.

Warner, S. L. (1965). RR: a survey technique for eliminating evasive answer bias. *J. Amer. Statist. Assoc.* **60**, 63–69.

Winkler, R. L., and Franklin, L. A. (1979). Warner's RR model: a Bayesian approach. *J. Amer. Statist. Assoc.* **74**, 207–214.

7

RR in a Finite Population Setting: A Unified Approach; Sampling with Varying Probabilities

7.1. INTRODUCTION

In the preceding chapters we described theories and methods of RR virtually, rather effectively, with reference to hypothetical populations. So we could allow population proportions bearing specific attributes to take up values continuously over the interval [0,1]. Moreover, only simple random sampling with replacement was permitted throughout so far with a minor exception in Appendixes 1 and 2, where selection could be without replacement. But if we consider applying RR to a finite population of N individuals, the proportions can take up only $(N+1)$ values, namely, i/N $(i=0, 1, \ldots, N)$. Similarly, a quantitative character may appear as the fixed values $Y_i (i=1, \ldots, N)$ of a variable y defined on the N individuals labeled as $i=1, \ldots, N$, instead of being treated directly as a random variable. As in usual survey sampling with open queries, one may use the labels to choose the samples with arbitrary but appropriately different probabilities. We will now treat the application of RR in a strictly finite population setup characterized by "identifiability" of units when samples are available according to suitable designs. A design, in particular a sampling design, prescribes selection probabilities to be assigned to various samples which are combinations of various labels out of $I_N=(1, \ldots, i, \ldots, N)$.

7.2. LINEAR UNBIASED ESTIMATORS

Let U be a finite population consisting of N units labeled $1, \ldots, i, \ldots, N$ with unknown values $Y_1, \ldots, Y_i, \ldots, Y_N$ of a sensitive variable y. For example, y may denote the amount of tax evaded, the extent of illegal and clandestine earnings, expenses for gambling, excessive alcoholic consumption, and similar items which people usually wish to conceal. The problem is to estimate the total $Y = \Sigma_1^N Y_i$ or the mean $\bar{Y} = Y/N$ when Y_i's are considered directly unavailable through open queries. For this a typical sample s from U may be chosen with a probability $p(s)$ corresponding to a design p. In the case of an open survey when y is innocuous and the respondents have no objection to give out their true values, one may employ a linear unbiased estimator for Y of the form

$$e = e(\underline{Y}) = e(s, \underline{Y}) = a_s + \sum_{i \in s} b_{si} Y_i \qquad (7.1)$$

such that $E_p e(s, \underline{Y}) = Y$, writing E_p for the expectation operator with respect to p. Here $\underline{Y} = (Y_1, \ldots, Y_i, \ldots, Y_N)$ and $e(s, \underline{Y})$ is free of Y_i's for $i \notin s$ but may involve Y_i's for $i \in s$. Also, a_s and b_{si}'s are free of $\underline{Y}$ and are subject to

$$\sum_s a_s p(s) = 0 \qquad \sum_{s \supset i} b_{si} p(s) = 1 \qquad (i = 1, \ldots, N) \qquad (7.2)$$

Here Σ_s is the sum over all samples and $\Sigma_{s \supset i}$ denotes the sum over samples that contain i. Suppose, instead, that y is sensitive and that most people are unlikely to divulge their y values. So RR trials are necessary to generate appropriate data. We will cite certain requisite procedures and present modified versions of the linear unbiased estimator (LUE) $e(\underline{Y})$ in (7.1).

Eriksson (1973a, b) initiated the following program. For each label i selected, he recommends using a randomization device independently $k(>1)$ times each to produce random observations Z_{ir}, where for $r = 1, \ldots, k$,

$$\begin{aligned} Z_{ir} &= Y_i && \text{with probability } C \\ &= X_j && \text{with probability } q_j \qquad (j = 1, \ldots, M) \end{aligned} \qquad (7.3)$$

Here $X_1, \ldots, X_M$ are predetermined real numbers supposed to cover the anticipated range of the Y_i's $(i = 1, \ldots, N)$. The quantities q_j $(j = 1, \ldots, M)$ are suitably chosen nonnegative proper fractions and C is a rightly assigned positive proper fraction such that $C + \Sigma_{j=1}^M q_j = 1$. A corresponding device is independently employed for every individual in the sample so that the values Z_{ir}, $r = 1, \ldots, k$, for $i \in s$ are generated. For the sake of a theoretical development the random vectors $\underline{Z}_r = (Z_{1r}, \ldots, Z_{ir}, \ldots, Z_{Nr})$, $r = 1, \ldots, k$, are conceptually supposed to be defined for every $i = 1, \ldots, N$. Let $\underline{Z} = (\underline{Z}_1, \ldots, \underline{Z}_k)$. In what follows we will write E_R ($V_{R'}$, C_R) to denote the operator for taking expectation (variance, covariance) with respect to the specific randomization technique employed to yield RRs (e.g., in this case the Z_{ir}'s). Further, writing

$$\bar{Z}_i=\frac{\sum_{r=1}^{k} Z_{ir}}{k} \qquad \mu_X=\frac{\sum_{j=1}^{M} q_j X_j}{1-C}$$

we note that

$$E_R(Z_{ir})=CY_i+\sum_{j=1}^{M} q_j X_j=CY_i+(1-C)\mu_X \tag{7.4}$$

and $E_R(\bar{Z}_i)=CY_i+(1-C)\mu_X$, $i=1, \ldots, N$ and $r=1, \ldots, k$. So Eriksson (1973b) suggests estimating Y_i by

$$\hat{Y}_i=\frac{\bar{Z}_i-(1-C)\mu_X}{C} \tag{7.5}$$

and then using the linear estimator (LE) for Y, namely,

$$e(\underline{Z})=e(s,\underline{Z})=a_s+\sum_{i\in s} b_{si}\hat{Y}_i \tag{7.6}$$

which is a version of (7.1) involving $\underline{Z}$ only through the Z_{ir}'s actually observed for $i\in s$, $r=1, \ldots, k$.

By $E_p(V_p, C_p)$ we will denote the operator for expectation (variance, covariance) with respect to the design p. Also, we will write $E_{pR}=E_p E_R$ $=E_R E_p=E_{Rp}$, $V_{pR}=V_{Rp}$ to indicate operators for expectation and variance, respectively, with respect to randomization followed by sampling design, or vice versa, assuming commutativity.

It follows that

$$E_R e(s,\underline{Z})=a_s+\sum_{i\in s} b_{si}Y_i=e(s,\underline{Y})$$

and that

$$E_{pR}e(s,\underline{Z})=\underline{Y} \tag{7.7}$$

The relation (7.7) will be interpreted to mean that $e(\underline{Z})=e(s,\underline{Z})$ is an LUE for Y in the RR setup in this context. We may say that $e(\underline{Z})$ is pR unbiased or "design-randomization unbiased" or just unbiased in brief. The variance of $e=e(s,\underline{Z})$ will be taken as

$$V_{pR}(e)=V_p[E_R(e)]+E_p[V_R(e)] \tag{7.8}$$

Writing

$$\sigma_X^2=\frac{\sum_{j=1}^{M} q_j(X_j-\mu_X)^2}{(1-C)} \qquad \text{and} \qquad \sigma_i^2=\sigma_X^2+C(Y_i-\mu_X)^2$$

it follows that

$$V_R(Z_{ir})=(1-C)[\sigma_X^2+C(Y_i-\mu_X)^2]=(1-C)\sigma_i^2 \qquad (i=1, \ldots, N;\ r=1, \ldots, k)$$

Then one gets

$$V_R(e) = \frac{\sum_{i \in s} b_{si}^2 V_R(Z_{ir})}{kC^2}$$

$$= \frac{1-C}{kC^2} \sum_{i \in s} b_{si}^2 \sigma_i^2$$

and hence

$$V_{pR}(e) = V_p[e(s, \underline{Y})] + \frac{1-C}{kC^2} \sum_1^N \sigma_i^2 \sum_{s \supset i} b_{si}^2 p(s) \tag{7.9}$$

This formula bears out how efficiency is lost when one uses RRs rather than DRs in case the latter are available from the sampled persons.

If in $e(s, \underline{Y})$ and hence in $e(s, \underline{Z})$ we take $a(s) = 0$ [i.e. we use a homogeneous linear unbiased estimator (HLUE)], then a simple estimator for the variance of $e(s, \underline{Z})$ is available as follows. In this case the design variance of $e(s, \underline{Y})$ may be written in the form

$$V_p[e(s, \underline{Y})] = \sum_1^N c_i Y_i^2 + \sum\sum_{i \neq j}^N d_{ij} Y_i Y_j \qquad \text{say}$$

A design-unbiased estimator for this is available provided we assume that the design p is so employed that the inclusion probabilities of units $\pi_i = \Sigma_{s \supset i} p(s)$ and of paired units $\pi_{ij} = \Sigma_{s \supset i,j} p(s)$ are all positive valued. Suppose that such an unbiased estimator (UE) is

$$v(s, \underline{Y}) = \sum_{i \in s} f_{si} Y_i^2 + \sum\sum_{i \neq j \in s} g_{sij} Y_i Y_j$$

such that $E_p[v(s, \underline{Y})] = V_p[e(s, \underline{Y})]$, with f_{si}'s and g_{sij}'s as quantities free of $\underline{Y}$. Let

$$\phi_i = \frac{\sum_{r=1}^k Z_{ir}^2/k - (1-C)(\sigma_X^2 + \mu_X^2)}{C}$$

Then $E_R(\phi_i) = Y_i^2$. Also, $E_R(\hat{Y}_i \hat{Y}_j) = Y_i Y_j$ for every $i \neq j$ in s. Hence defining

$$v(s, \underline{Z}) = \sum_{i \in s} f_{si} \phi_i + \sum\sum_{i \neq j \in s} g_{ijs} \hat{Y}_i \hat{Y}_j$$

one obtains

$$E_{pR}[v(s, \underline{Z})] = E_p[E_R v(s, \underline{Z})] = E_p[v(s, \underline{Y})] = V_p[e(s, \underline{Y})] \tag{7.10}$$

Let

$$s_{Zi}^2 = \frac{\sum_{r=1}^k (Z_{ir} - \bar{Z}_i)^2}{k-1} \qquad (i \in s)$$

Then

$$E_{\mathrm{R}}(s_{Zi}^2)=V_{\mathrm{R}}(Z_{ir}) \qquad (r=1, \ldots, k)$$

$$E_{\mathrm{R}}\frac{\sum_{i\in s} b_{si}^2 s_{Zi}^2}{kC^2}=\frac{\sum_{i\in s} b_{si}^2 V_{\mathrm{R}}(Z_{ir})}{kC^2}=V_{\mathrm{R}}(e)$$

and

$$E_{p\mathrm{R}}\frac{\sum_{i\in s} b_{si}^2 s_{Zi}^2}{kC^2}=E_p[V_{\mathrm{R}}(e)] \tag{7.11}$$

Using (7.7) to (7.11), one may check that

$$E_{p\mathrm{R}}\left[v(s, \underline{Z})+\frac{\sum_{i\in s} b_{si}^2 s_{Zi}^2}{kC^2}\right]=V_{p\mathrm{R}}(e)$$

So $\hat{v}(e)=v(s, \underline{Z})+\Sigma_{i\in s} b_{si}^2 s_{Zi}^2/kC^2$ may be taken as a UE for $V_{p\mathrm{R}}(e)$. Unbiasedness of estimators based on RR data is naturally defined in terms of the operator $E_{p\mathrm{R}}$.

The estimator $e(s, \underline{Z})$ is proposed as a counterpart of $e(s, \underline{Y})$. Except for its unbiasedness in terms of the operator $E_{p\mathrm{R}}$ and the existence of an unbiased variance estimator $\hat{v}(e)$ for $V_{p\mathrm{R}}(e(s, \underline{Z}))$, no other property, optimal or otherwise, is known or investigated by Eriksson (1973a,b).

Godambe (1980) was the first to investigate optimal sampling strategies in the RR setup. He envisages a situation where, in addition to the sensitive variate y of immediate interest, there is an innocuous auxiliary variate, say, w with unknown values $W_1, \ldots, W_N$ for the N units of U supposed to be closely related to y. But he assumes knowledge of the total $W=\Sigma_1^N W_i$. For every individual in a sample s chosen according to a design p as before, a suitable randomizing device is independently employed. According to such a device, let us have, for an individual labeled i, the randomized response Z_i, where for a suitably chosen C, a positive proper fraction,

$$\begin{aligned} Z_i &= Y_i, && \text{with probability } C \\ &= W_i && \text{with probability } 1-C \end{aligned} \tag{7.12}$$

Although Z_i is available actually only for i in s, for the sake of theoretical necessity we assume that the vector

$$\underline{Z}=(Z_1, \ldots, Z_i, \ldots, Z_N)$$

is defined, as in the case of Eriksson's setup. Godambe (1980) considers using an LUE based on the RR data $(s, Z_i | i\in s)$ of the form

$$e=e(\underline{Z})=e(s, \underline{Z})=a_s+\sum_{i\in s} b_{si} Z_i \tag{7.13}$$

with a_s and b_{si} as real numbers free of Z_i and W_i, so that $e(s, \underline{Z})$ is free of Z_i for $i \notin s$. He calls it unbiased for $Y = \Sigma_i^N Y_i$ when, as earlier, the following holds:

$$E_{p\mathrm{R}}[a_s + \textstyle\sum_{i \in s} b_{si} Z_i] = Y$$

Since by (7.12) one has

$$E_{\mathrm{R}}(Z_i) = CY_i + (1-C)W_i = \theta_i$$

say, a necessary and sufficient condition on p, a_s, and b_{si} for unbiasedness of e is that

$$\textstyle\sum_s a_s p(s) + \dfrac{W(1-C)}{C} = 0$$

$$\textstyle\sum_{s \supset i} b_{si}\, p(s) = \dfrac{1}{C} \qquad \forall \quad i = 1, \ldots, N \qquad (7.14)$$

While Eriksson (1973b) does not introduce any criterion for discriminating among his LUEs for Y, Godambe (1980) makes some studies in this direction. The variance of e is obviously

$$V_{p\mathrm{R}}(e) = E_p[V_{\mathrm{R}}(e)] + V_p[E_{\mathrm{R}}(e)] = E_{\mathrm{R}}[V_p(e)] + V_{\mathrm{R}}[E_p(e)]$$

This variance can become minimal if a_s and b_{si} are chosen subject to (7.14) such that the quantity $E_{\mathrm{R}} E_p(e^2)$ is minimized. To make such choices for a_s and b_{si}, introducing Lagrangian undetermined multipliers λ and μ_i, $i = 1, \ldots, N$, one has to solve the equations

$$\frac{\partial}{\partial a_s}\left\{ E_{\mathrm{R}} E_p e^2(s, \underline{Z}) - \lambda\left[\textstyle\sum a_s p(s) + \dfrac{W(1-C)}{C}\right] - \textstyle\sum_{i=1}^N \mu_i \left[\textstyle\sum\limits_{s \supset i} b_{si} p(s) - \dfrac{1}{C}\right]\right\} = 0$$

$$\frac{\partial}{\partial b_{si}}\left\{ E_{\mathrm{R}} E_p e^2(s, \underline{Z}) - \lambda\left[\textstyle\sum a_s p(s) + \dfrac{W(1-C)}{C}\right] - \textstyle\sum_1^N \mu_i \left[\textstyle\sum\limits_{s \supset i} b_{si} p(s) - \dfrac{1}{C}\right]\right\} = 0$$

These, respectively, lead to the equations

$$a_s + \textstyle\sum_{i \in s} b_{si}\, \theta_i = \dfrac{\lambda}{2} \qquad (7.15)$$

$$E_{\mathrm{R}}\left(\textstyle\sum_{i \in s} b_{si} Z_i\right) Z_i = \dfrac{\mu_i}{2} - a_s \theta_i \qquad (7.16)$$

for every i in s with $p(s)>0$. When combined, (7.15) and (7.16) lead to

$$E_{\mathrm{R}}\left[\left(\sum_{i\in s} b_{si}Z_i\right)Z_i-\left(\sum_{i\in s} b_{si}\theta_i\right)\theta_i\right]=\frac{1}{2}(\mu_i-\lambda\theta_i) \tag{7.17}$$

Writing $V_{\mathrm{R}}(Z_i)=V_i$, this leads to

$$b_{si}=\frac{\mu_i-\lambda\theta_i}{2V_i}$$

for every i in s with $p(s)>0$ and $V_i>0$. This means that b_{si} must be of the form b_i, free of s, for every s such that $s \supset i$ and $p(s)>0$. By (7.14) it follows that the optimal b_{si}, say $\bar{b}_{si}$, must equal $1/C\pi_i$. But with this $\bar{b}_{si}$, using (7.14) together with (7.15), it follows that the optimal choice of a_s as, say, $\bar{a}_s$ is impossible because $\bar{a}_s$ involves $\underline{Y}$. To get over this, one possibility is to take $a_s=0$ in $e(s, \underline{Z})$ itself (i.e., to confine to the use of HLUEs). If this is conceded, then also, as is easy to check, the optimal choice of b_{si} is $\bar{b}_{si}=1/C\pi_i$ as earlier, but subject to the condition

$$E_{\mathrm{R}}Z_i\left(\sum_{i\in s}\frac{Z_i}{C\pi_i}\right)=\frac{\mu_i}{2} \tag{7.18}$$

This, on simplification, is expressible as

$$\sum_{i\in s}\frac{\theta_i}{\pi_i}=\frac{C\mu_i/2-V_i/\pi_i}{\pi_i}=\Psi_i \quad \text{(say)} \tag{7.19}$$

But for every s with $p(s)>0$ such that $s \supset i$, the relation (7.19) cannot hold for an arbitrary design p with $\pi_i>0$ except when it is a unicluster design. This is similar to that in the case of open surveys, as is well known from the works of Godambe (1955, 1965), Hege (1965), Hanurav (1966), and Lanke (1975).

Because of the nonexistence of a UMV estimator (UMVE) in general, a problem is how to discriminate among competing LUEs or HLUEs. In developing a criterion for comparison, Godambe (1980) postulates a superpopulation model (M, say) so as to treat $\underline{Y}$ as a random vector with independently distributed coordinates. By $E_m(V_m, C_m)$ we will denote the operator for expectation (variance, covariance) with respect to the distribution of $\underline{Y}$. Also, let p_n denote generically a design such that only samples with a fixed effective size n have positive probabilities for selection. By $\bar{p}_n$ we will mean, generically, a design assigning positive selection probabilities only to those samples that have n as the average effective size in the sense that $\Sigma_s\, \nu(s)p(s)=n$, denoting by $\nu(s)$ the number of distinct units in a sample s

with selection probability $p(s)$. By $\mathscr{L}$ we will mean the class of LUEs for Y of the form (7.13) subject to (7.14). Godambe (1980) suggests assessing the performances of sampling strategies (p,e) in terms of the characteristics $E_m V_{pR}(e)$ keeping e in the class $\mathscr{L}$ and taking for p a p_n design. An optimal subclass of strategies is then derived by Godambe (1980), as we note below in brief.

We will assume that E_m commutes with E_p and E_R and write $E_m(\theta_i)=\xi_i$, $\underline{\boldsymbol{\theta}}=(\theta_1, \ldots, \theta_N)$, $E_m(Y_i)=\zeta_i$, $V_m(\theta_i)=D_i$, $\theta=\Sigma\theta_i$, $\xi=\Sigma\xi_i$, $\zeta=\Sigma\zeta_i$, $e(s, \underline{\boldsymbol{\theta}})=a_s+\Sigma_{i\in s}b_{si}\theta_i$, and $\Delta_m(e)=E_m[e(s, \underline{\theta})-E_m(Y)]$. Then we have

$$\begin{aligned} E_m V_{pR}(e) &= E_m E_p E_R[e(s, \underline{Z})-Y]^2 \\ &= E_m E_p E_R\{[e(s, \underline{Z})-E_R e(s, \underline{Z})]+[E_R e(s, \underline{Z})-Y]\}^2 \\ &= E_m E_p\left(\sum\nolimits_{i\in s} b_{si}^2 V_i\right)+E_m V_p\left(a_s+\sum\nolimits_{i\in s} b_{si}\theta_i\right) \end{aligned}$$

Now

$$E_p\sum\nolimits_{i\in s} b_{si}^2 V_i=\sum\nolimits_{i=1}^{N} V_i\left[\sum\nolimits_{s\supset i} b_{si}^2 p(s)\right]\geq\sum\frac{V_i}{C^2\pi_i}$$

and the lower bound is attained if b_{si} is taken as $1/C\pi_i$ for every s with $p(s)>0$. Also,

$$\begin{aligned} &E_m V_p\left(a_s+\sum\nolimits_{i\in s} b_{si}\theta_i\right) \\ &\quad=E_m E_p[e(s, \underline{\boldsymbol{\theta}})-Y]^2 \\ &\quad=E_m E_p\{[e(s, \underline{\boldsymbol{\theta}})-E_m e(s, \underline{\boldsymbol{\theta}})]+[E_m e(s, \underline{\boldsymbol{\theta}})-E_m(Y)]-[Y-E_m(Y)]\}^2 \\ &\quad=E_p V_m[e(s, \underline{\boldsymbol{\theta}})]+E_p\Delta_m^2(e)-V_m(Y) \\ &\quad=\sum\nolimits_i D_i\sum\nolimits_{s\supset i} b_{si}^2 p(s)+E_p\left(a_s+\sum\nolimits_{i\in s} b_{si}\xi_i-\zeta\right)^2-V_m(Y) \end{aligned}$$

This quantity is minimized for the choice $b_{si}=1/C\pi_i$, $a_s=-(1-C)/C\Sigma_{i\in s}(W_i/\pi_i)$, provided that one has a design p as a p_n with

$$\pi_i=\frac{n\zeta_i}{\zeta} \qquad (i=1, \ldots, N) \tag{7.20}$$

A p_n design with $\pi_i=n\zeta_i/\zeta\ \forall\ i$ will be denoted as $p_n(\zeta)$. The estimator

$$\sum\nolimits_{i\in s}\frac{Z_i}{C\pi_i}-\frac{1-C}{C}\sum\nolimits_{i\in s}\frac{W_i}{\pi_i}$$

will be denoted as $\bar{e}=\bar{e}(s, \underset{\sim}{Z})$. By $(p_n(\zeta), \bar{e})$ we will denote a subclass of strategies out of the class (p_n, e) with e in $\mathscr{L}$. The results due to Godambe (1980) may now be stated as two theorems.

Theorem 7.1

(a) Among estimators $e(s, \underset{\sim}{Z})=a_s+\Sigma_{i\in s}b_{si}Z_i$ with a_s's and b_{si}'s free of $\underset{\sim}{Z}$ subject to (7.14), there does not exist one with the least value of $E_{p\mathrm{R}}[e(s, \underset{\sim}{Z})-Y]^2$, whatever fixed design p one may employ.

(b) Among estimators $e(s, \underset{\sim}{Z})=\Sigma_{i\in s}b_{si}Z_i$ with b_{si}'s free of $\underset{\sim}{Z}$ subject to $\Sigma_{s\ni i}b_{si}p(s)=1/C$ based on any nonunicluster design there does not exist one with the least value of $E_{p\mathrm{R}}\ (\Sigma_{i\in s}b_{si}Z_i-Y)^2$.

Theorem 7.2. Under a super population model stipulating $\underset{\sim}{Y}$ as a random vector with Y_i's distributed independently with $E_m(Y_i)=\zeta_i$, $i=1, \ldots, N$, among all strategies (p_n, e) with e in $\mathscr{L}$ an optimal subclass is $(p_n(\zeta), \bar{e})$, in the sense that

$$E_m V_{p_n\mathrm{R}}(e)\geq E_m V_{p_n(\zeta)\mathrm{R}}(\bar{e})$$

Remark. Since the W_i's are unknown, and ζ_i's should also be unavailable in practice a strategy $(P_n(\zeta), \bar{e})$ is not usable. To resolve this impasse, Godambe (1980) suggested certain approximations. One does not, however, know the status of a strategy derived through his approximations. Since the W_i's are not known a priori but only W is assumed given, his approximate estimator is

$$\bar{e}'=\frac{N}{n}\sum_{i\in s}\frac{Z_i}{C}-\frac{(1-C)W}{C}$$

His simplified design is a p_n with $\pi_i=n/N$, $i=1, \ldots, N$. Such an equal probability sampling is advised because ζ_i's are unlikely to be known in practice. But to promote the level of efficiency, Godambe (1980) also recommends stratification if some appropriate guess may be made about the Y_i's. For further details a curious reader may consult the original work of Godambe (1980).

7.3. LINEAR ESTIMATION WITH RR SUBJECT TO OBSERVATIONAL ERRORS

Bellhouse (1980) considers estimating $\bar{Y}$ on taking a sample s with probability $p(s)$ and extracting through a random device a quantitative RR as Z_i from a sampled individual labeled i. He does not specify his RR device but stipulates that the expected response $E_R(Z_i)$ should equal a number Y_i^* (say). This Y_i^* is supposed to be an estimate of Y_i but subject to random error of observation due to spurious reporting and/or recording. Using the symbol E_r to denote the operator for expectation with respect to the distribution of random reporting/recording error, he supposes that $E_r(Y_i^*)=Y_i$ for every $i=1, \ldots, N$. He also considers LUEs of the form (7.13). He immediately postulates a superpopulation model and demands his estimator of the form $e=e(s, \underline{Z})=a_s+\Sigma_{i\in s}b_{si}Z_i$ to satisfy the condition

$$E_m E_p E_r E_R\left(a_s+\sum\nolimits_{i\in s}b_{si}Z_i\right)=E_m(\bar{Y})=\bar{\mu} \quad \text{say} \tag{7.21}$$

Such an estimator e subject to (7.21) is called a *prm*-unbiased or model-design-unbiased estimator. The class of estimators (7.13) subject to (7.21) will be denoted by $\mathscr{B}$. Here e is directly an estimator for $\bar{\mu}$ and a predictor for $\bar{Y}$.

Bellhouse (1980) considers only fixed (at n)-size designs p_n and considers the following notations:

W_i's are certain known positive weights, $r_i=Y_i/W_i$, $\bar{R}=\Sigma r_i/N$, $\bar{W}=\Sigma W_i/N$, α, β are known real constants. He assumes r_i's and W_i's to be uncorrelated and hence is motivated to postulate that

$$\frac{Z_i}{W_i}=\alpha+\beta\bar{R}+\varepsilon_i \qquad (i=1, \ldots, N)$$

$$\mathscr{E}(\varepsilon_i)=0 \qquad \mathscr{E}(\varepsilon_i^2)=\gamma_1(>0)$$

$$\mathscr{E}(\varepsilon_i\varepsilon_j)=\gamma_2 \qquad (i\neq j)$$

using the operator $\mathscr{E}=E_m E_r E_R$. Writing $E_n\equiv E_m E_{p_n} E_r E_R$ with each expectation operator commuting with every other, he shows that among all strategies (p_n, e) with e in $\mathscr{B}$, an optimal subclass is $(p_n, \hat{\bar{Y}})$, where

$$\hat{\bar{Y}}=\frac{\bar{W}(1/n)\sum_{i\in s}(Z_i/W_i)-\alpha}{\beta} \tag{7.22}$$

in the sense that for every e in $\mathscr{B}$,

$$E_n(e-\bar{\mu})^2\geq E_n(\hat{\bar{Y}}-\bar{\mu})^2=\frac{\bar{W}^2[\gamma_1+(n-1)\gamma_2]}{n\beta^2} \tag{7.23}$$

A proof of (7.23) follows on appealing to a theorem of Rao (1973), which in the present context may be rephrased to have the following version.

Theorem 7.3. Among all *pm*-unbiased nonhomogeneous linear estimators e for $\bar{\mu}$ in $\mathscr{B}$, the optimal one in the sense of having the minimal value for $E_n(e-\bar{\mu})^2$ is, say, $\bar{e}$ such that $E_n(\bar{e}e_0)=0$ for every e_0 in $\mathscr{B}_0$, where $\mathscr{B}_0$ is a subclass of $\mathscr{B}$ with the restriction that $\bar{\mu}=0$, bearing in mind that only p_n (with any preassigned integer n) designs are in competition.

To prove the relation (7.23), one requires only to verify that if

$$e_0=d_s+\sum_{i\in s}d_{si}Z_i$$

with d_s and d_{si} free of $\underline{Z}$ is such that $E_n(e_0)=0$, then $E_n(\hat{\hat{\bar{Y}}}e_0)=0$. Verification of this is simple on taking any p_n and hence is not demonstrated here to save space and is left as an exercise for the reader.

It is interesting to examine the fate of the estimator $\hat{\hat{\bar{Y}}}$ when one restricts within the class of design-unbiased nonhomogeneous linear estimators $e=b_s+\Sigma_{i\in s}b_{si}Z_i$ for $\bar{Y}$. By design unbiasedness we mean that $E_pE_rE_R(e)=\bar{Y}$. In this case $\underset{\sim}{e}$ is an estimator for $\bar{Y}$.

It is easy to check that $\hat{\hat{\bar{Y}}}$ is design-unbiased provided that (1) it is based on a p_n design for which $\pi_i=\mathrm{n}W_i/N\bar{W}$, $i=1,\ldots,N$, and (2) an RR technique is so employed that $E_rE_R(\bar{Z})=\alpha\bar{W}+\beta\bar{Y}$. We may denote any such design as $p_n(W)$.

Let e be p-unbiased for $\bar{Y}$. Following the analysis in Section 7.2, we may note, on writing $\mathscr{V}$ as the variance operator corresponding to the expectation operator $\mathscr{E}$, that

$$\begin{aligned}E_n(e_b-\bar{Y})^2&=\mathscr{E}E_{p_n}(e_b-\bar{Y})^2=E_{p_n}\mathscr{V}(e_b)+E_{p_n}[\mathscr{E}(e_b-\bar{Y})]^2-V_m(\bar{Y})\\&\geq E_{p_n}(W)\mathscr{V}(\hat{\hat{\bar{Y}}})-V_m(\bar{Y})=E_n(\hat{\hat{\bar{Y}}}-\bar{Y})^2 \qquad (7.24)\end{aligned}$$

when $\hat{\hat{\bar{Y}}}$ is based on a $p_n(W)$ design, on observing that $\mathscr{E}(\hat{\hat{\bar{Y}}}-\bar{Y})=0$. This optimality of $(p_n(W),\hat{\hat{\bar{Y}}})$ among (p_n,e), with e in $\mathscr{L}$ may also be checked from Theorem 2.4 in Rao and Bellhouse (1978), as noted by Bellhouse (1980). Bellhouse (1980) illustrates practical applications of optimality results implied by (7.23) and (7.24) as we describe briefly below.

Example 7.1. Suppose that a $p_n(W)$ design is employed. Following Pollock and Bek's (1976) procedure, let a randomizing device be employed by a chosen individual labeled i to get a value X_i on a random variable x. Further, let him or her respond a value $Z_i=Y_i+W_iX_i$. Let $E_R(X_i)=\alpha$, for every i, $V_R(X_i)=\sigma_x^2$, $C_R(X_i,X_j)=0$, $i\neq j$. Let $\underline{Y}$ be so modeled that $\underline{r}=(r_1,\ldots r_i,\ldots,r_N)$ is a random realization of one of the $N!$ vectors $(r_{j_1},\ldots r_{j_N})$, where $(j_1,\ldots j_N)$ is a permutation of $I_N=(1,\ldots,N)$. Under this random permutation model, denoted as M_{Rp}, say it follows that $E_mE_R(Z_i/W_i)=\bar{R}+\alpha$. Here reporting/recording error is ignored. Then this is a special case of the Bellhouse model with $\beta=1$, $\gamma_1=\sigma_r^2+\sigma_x^2$, and $\gamma_2=-\sigma_r^2/(N-1)$, where $\sigma_r^2=\Sigma(r_i-\bar{R})^2/N$. So the optimal *prm*-unbiased estimator in $\mathscr{B}$ for $\bar{\mu}$ is $\bar{W}\,[(1/n)\,\Sigma_{i\in s}(Z_i/W_i)-\alpha]$. This is also the optimal design-unbiased estimator for $\bar{Y}$.

Example 7.2. Let y be a sensitive variable of interest and x an unrelated innocuous variate. A simple random sample of size n is supposed to be taken without replacement. Each selected person labeled, say i, is asked, with a probability P, to report his or her value of y, available possibly with a random reporting error, as, say, Y_i^*. But he or she is asked, with a probability $(1-P)$ to report the value of x available, with no reporting error, as, say, X_i. This is a version of Greenberg *et al.*'s (1969) unrelated-question model. The resulting RR denoted as Z_i elicited from the ith sample person has $E_R(Z_i)=PY_i^*+(1-P)X_i$, $E_rE_R(Z_i)=PY_i+(1-P)X_i$. Further, $\underline{X}=(X_1,\ldots,X_N)$, $\underline{Y}=(Y_1,\ldots,Y_N)$ are so modeled that

$$((X_1,Y_1),\ldots,(X_i,Y_i),\ldots,(X_N,Y_N))$$

is regarded as a random realization of one of the $N!$ vectors $((X_{j_1},Y_{j_1}),\ldots(X_{j_N},Y_{j_N}))$, with X_i's and Y_i's individually unknown but $\bar{X}=\Sigma X_i/N$ is known and $(j_1,\ldots,j_N)$ is a permutation of $(1,\ldots,N)$. This random permutation model M_{Rp} (Rao and Bellhouse, 1978) turns out a special case of the Bellhouse model with

$$W_i=1 \qquad \beta=P \qquad \alpha=(1-P)\bar{X}$$

If reporting/recording errors are absent (i.e., $Y_i^*=Y_i$), then with $\sigma_{uv}=(1/N)\Sigma[u_i-(1/N)\Sigma u_i]\,[v_i-(1/N)\Sigma v_i]$ $(u,v=x,y)$, one has

$$\gamma_1=P\sigma_{yy}+(1-P)\sigma_{xx}+P(1-P)(\bar{X}-\bar{Y})^2$$

$$\gamma_2=-\frac{P^2\sigma_{yy}+2P(1-P)\sigma_{xy}+(1-P)^2\sigma_{xx}}{N-1}$$

It follows that the optimal design-unbiased estimator for $\bar{Y}$, and also optimal model-design-unbiased predictor for $\bar{Y}$ among those in $\mathscr{B}$, is

$$\hat{\bar{Y}}=\frac{\bar{z}-(1-P)\bar{X}}{P} \tag{7.25}$$

where $\bar{z}$ is the sample mean of RRs.

Remark. Eriksson (1973a,b), Godambe (1980), and Bellhouse (1980) restrict their investigations within the class of linear estimators based on randomized responses. Eriksson needs replicated RRs from each respondent to get a variance estimator, but he is silent about optimality properties, if any, of his strategies. On his own admission, Godambe's (1980) optimal strategy is not practicable. Bellhouse (1980) does not specify any particular mode of generating RRs but illustrates applicability of his optimality results with reference to some standard procedures already in vogue. In the next section we remove the linearity restriction on estimators and note additional optimality results.

7.4. OPTIMALITY OF GENERAL UNBIASED ESTIMATORS

As before, let p be any design and s be a sample to be chosen with probability $p(s)$. From respective individuals labeled $i(=1, \ldots, N)$, let independent RRs be obtainable as Z_i $(i=1, \ldots, N)$ such that

$$\begin{aligned} Z_i &= Y_i \quad \text{with probability } C \qquad 0<C<1 \\ &= X_j \quad \text{with probability } (1-C)q_j \end{aligned}$$

where $j=1, \ldots, M$, $\Sigma_{j=1}^{M} q_j=1$.

The X_j's are supposed to be sufficiently numerous chosen suitably so that they together coincide with the range of possible values of Y_i's when appropriately rounded off, in any particular context. As usual, let $\underline{Z} =(Z_1, \ldots, Z_i, \ldots, Z_N)$ be defined conceptually and C be appropriately assignable. Let $\pi_i=\Sigma_{s \ni i}\, p(s)>0$ and $\mu_X=\Sigma_{j=1}^{M} X_j q_j$.

For a given s, let $e(\underline{Y})=e(s,\underline{Y})$ be free of Y_i's for $i \notin s$ and $E_p e(s,\underline{Y})=Y$ for every $\underline{Y}$. Corresponding to this, let $e(\underline{Z})=e(s,\underline{Z})$ be a function of s and $\underline{Z}$ free of Z_i's for $i \notin s$ so that $E_p\, e(s,\underline{Z})=\Sigma_1 Z_i=Z$.

In what follows we will essentially recount the contents in the works of Adhikary *et al.* (1984) and Chaudhuri (1986). Let

$$t=t(s,\underline{Z})=\frac{e(s,\underline{Z})-(1-C)\mu_X \sum_{i\in s} 1/\pi_i}{C} \tag{7.26}$$

Then $E_p(t)=[Z-(1-C)N\mu_X]/C$ and

$$E_{pR}(t)=E_{Rp}(t)=E_R[E_p(t)]=Y$$

Thus t is pR unbiased for Y. It is a derived estimator corresponding to $e(s,\underline{Y})$ and is the randomized version of the latter available only in open surveys. We will note from Theorems 7.4 and 7.5 that under the assumptions and formulations above, any pR unbiased estimator for Y based on $(s, Z_i | i \in s)$, say $a(s,\underline{Z})$, is essentially of the form $t=t(s,\underline{Z})$ in (7.26).

Theorem 7.4. $E_{pR}h(s,\underline{Z}) = E_R E_p h(s,\underline{Z}) = 0$ for every $\underline{Y} \Rightarrow E_p h(s,\underline{Z}) = 0$ for every $\underline{Z}$.

Proof. Let $E_p h(s,\underline{Z}) = \phi(\underline{Z})$ and $E_R \phi(\underline{Z}) = 0$. In case $N=1$, we have $0 = E_R \phi(Z_1) = C\phi(Y_1) + (1-C)\Sigma_1^M q_j \phi(X_j)$. Then $\phi(Y_1) = -[(1-C)/C]$ $\Sigma_{j=1}^M q_j \phi(X_j) =$ a constant $= k$, say for every Y_1 in its range. Since the range of Y_1 coincides with the range of X_j's ($j=1, \ldots, M$), it follows that $k=0$ [i.e., $\phi(Y_1)=0$]. If the theorem is true for all $N=1,2, \ldots, m$, say, then in case $N=m+1$, one has, since Z_i's are independent over $i=1, \ldots, N$, as every individual independently applies the randomized device,

$$0 = E_R \phi(Z_1, \ldots, Z_m, Z_{m+1}) = C E_R \phi(Z_1, \ldots, Z_m, Y_{m+1})$$
$$+ (1-C)\Sigma_l^M q_j E_R \phi(Z_1, \ldots, Z_m, X_j)$$

for every $Z_1, \ldots, Z_m$. This implies, arguing as before, that for every Y_{m+1},

$$E_R \phi(Z_1, \ldots, Z_m, Y_{m+1}) = 0$$

But by hypothesis, for a fixed Y_{m+1},

$$E_R \phi(Z_1, \ldots, Z_m, Y_{m+1}) = 0$$

for every $Y_1, \ldots, Y_m$. This implies

$$\phi(Z_1, \ldots, Z_m, Y_{m+1}) = 0$$

Since this is true for every Y_{m+1}, $\phi(Z_1 \ldots, Z_m, Z_{m+1}) = 0$ for every $Y_1, \ldots, Y_m, Y_{m+1}$. The proof is thus completed by induction.

Theorem 7.5. $E_{pR}\, a(s,\underline{Z}) = Y$ for every $\underline{Y}$ implies that $a(s,\underline{Z})$ coincides essentially with $t(s,\underline{Z})$ in (7.26).

Proof. Since $E_p[E_R a(s,\underline{Z})] = Y$, it follows that $E_R\, a(s,\underline{Z})$ can be taken as an estimator $e(s,\underline{Y})$, which is p unbiased for Y and usable in case of an open survey. Hence one may derive from this $e(s,\underline{Y})$ the estimator $t(s,\underline{Z})$ of the form (7.26) in the RR situation.

On the other hand, since for the $t(s,\underline{Z})$ in (7.26), $E_{pR}[a(s,\underline{Z}) - t(s,\underline{Z})] = 0$, it follows, by Theorem 7.4, that

$$E_p a(s,\underline{Z}) = E_p t(s,\underline{Z}) = \frac{\sum_1^N Z_i - (1-C)N\mu_X}{C}$$

Hence

$$E_p\left[Ca(s,\underline{Z}) + (1-C)\mu_X \sum_{i\in s} \frac{1}{\pi_i}\right] = \sum_1^N Z_i$$

Since this is true for every $\underline{Z}$ and hence for every $\underline{Y}$, we have $Ca(s,\underline{Y})+(1-C)\mu_X \Sigma_{i\in s}(1/\pi_i)$ is p unbiased for Y. So for an open survey, one may take

$$e(s,\underline{Y})=Ca(s,\underline{Y})+(1-C)\mu_X \sum_{i\in s}\frac{1}{\pi_i}$$

as a p unbiased estimator for Y and hence get from it the derived estimator as

$$a(s,\underline{Z})=\frac{e(s,\underline{Z})-(1-C)\mu_X \sum_{i\in s}(1/\pi_i)}{C}=t(s,\underline{Z})$$

Among the estimators $a(s, \underline{Z})$ for Y, the one of our main interest will be that for which $V_{pR}\, a(s, \underline{Z})$ is the minimum for every $\underline{Y}$. But Theorem 7.6 shows the nonexistence of such a uniformly minimum variance (UMV) unbiased estimator.

Theorem 7.6. A UMV pR-unbiased estimator for Y does not exist.

Proof. Let $\underline{Y}_0=(Y_{10}, \ldots, Y_{i0}, \ldots, Y_{N0})$ be a fixed vector point in the range of $\underline{Y}$. Also, let

$$\begin{aligned} Z_{i0} &= Y_{i0} && \text{with probability } C \\ &= X_j && \text{with probability } q_j\ (j=1, \ldots, M) \\ \underline{Z}_0 &= (Z_{10}, \ldots, Z_{i0}, \ldots, Z_{N0}) \end{aligned}$$

If possible, let $a(s, \underline{Z})$ be the UMV pR-unbiased estimator for Y. Let us define the pR-unbiased estimator

$$a^*(s, \underline{Z})=a(s, \underline{Z})-a(s, \underline{Z}_0)+E_R\, a(s, \underline{Z}_0)$$

Then

$$\begin{aligned} D(\underline{Z}) &= V_{pR}\, a(s, \underline{Z})-V_{pR}\, a^*(s, \underline{Z}) \\ &= \{V_p[E_R\, a(s, \underline{Z})]-V_p[E_R\, a^*(s, \underline{Z})]\} \\ &\quad +E_p\{[V_R\, a(s, \underline{Z})]-[V_R\, a^*(s, \underline{Z})]\} \end{aligned}$$

When $\underline{Z}$ coincides with $\underline{Z}_0$, $D(\underline{Z})$ equals $D(\underline{Z}_0)=E_pV_R\, a(s, \underline{Z}_0) > 0$. So $a(s, \underline{Z})$ cannot have the uniformly smallest variance.

Next consider a particular estimator of the form $t=t(s,\underline{Z})$, namely, $t^*=t^*(s,\underline{Z})=(1/C)\Sigma_{i\in s}(Z_i^*/\pi_i)$, where $Z_i^*=-(1-C)\mu_X+Z_i$. This t^* may be looked upon as a randomized version of the well-known Horvitz-Thompson (1952) estimator $e^*(s, \underline{Y})=\Sigma_{i\in s}(Y_i/\pi_i)$ in the open survey. Obviously, this is pR unbiased for Y. For any pR-unbiased estimator $e(s,\underline{Z})$ let us write

$$t(s,\underline{Z})=t^*(s,\underline{Z})+h(s,\underline{Z})$$

with $h=h(s,\underset{\sim}{Z})$ free of Z_i for $i\notin s$ when $p(s)>0$. Then it follows that $E_{pR}\ h(s,\underset{\sim}{Z})=0$, hence

$$E_p h(s,\underset{\sim}{Z})=0 \tag{7.27}$$

by Theorem 7.4. Hence

$$\sum\nolimits_{s\supset i} h(s,\underset{\sim}{Z})p(s)=-\sum\nolimits_{s\not\supset i} h(s,\underset{\sim}{Z})p(s) \tag{7.28}$$

$$V_{pR}t(s,\underset{\sim}{Z})=V_p[E_R t(s,\underset{\sim}{Z})]+E_p[V_R t(s,\underset{\sim}{Z})]$$

$$\begin{aligned}E_p V_R[t(s,\underset{\sim}{Z})]&=E_p V_R t^*(s,\underset{\sim}{Z})+E_p V_R h(s,\underset{\sim}{Z})\\&\quad+2E_p C_R[t^*(s,\underset{\sim}{Z}),h(s,\underset{\sim}{Z})]\end{aligned} \tag{7.29}$$

But using (7.28),

$$\begin{aligned}&E_p C_R(t^*(s,\underset{\sim}{Z}),h(s,\underset{\sim}{Z}))\\&=E_p E_R\left[\sum\nolimits_{i\in s}\frac{Z_i-E_R(Z_i)}{C\pi_i}\right]h(s,\underset{\sim}{Z})\\&=E_R\left[\sum\nolimits_{i=1}^{N}\frac{Z_i-E_R(Z_i)}{C\pi_i}\sum_{s\supset i}h(s,\underset{\sim}{Z})\,p(s)\right]\\&=-E_R\sum\nolimits_1^N\frac{Z_i-E_R(Z_i)}{C\pi_i}\sum_{s\not\supset i}h(s,\underset{\sim}{Z})\,p(s)\\&=0\end{aligned} \tag{7.30}$$

since Z_i's are independent. Using (7.29) and (7.30),

$$E_p V_R[e(s,\underset{\sim}{Z})]>E_p V_R[t^*(s,\underset{\sim}{Z})]$$

$$\begin{aligned}&V_{pR}[e(s,\underset{\sim}{Z})]-V_{pR}[t^*(s,\underset{\sim}{Z})]\\&\qquad>V_p[E_R\,e(s,\underset{\sim}{Z})]-V_p[E_R\,t^*(s,\underset{\sim}{Z})]\\&\qquad=V_p[E_R\,e(s,\underset{\sim}{Z})]-V_p[e^*(s,\underset{\sim}{Y})]\end{aligned} \tag{7.31}$$

The above leads to the following observations. For any pR-unbiased estimator $e(s,\underset{\sim}{Z})$ for Y based on the RR-survey data $(s, Z_i|i\in s)$, one may consider for Y the p-unbiased estimator $E_R\,e(s,\underset{\sim}{Z})$ based on open or direct survey data $(s, Y_i|i\in s)$. The latter may be treated as a possible competitor for $e^*(s,\underset{\sim}{Y})=\Sigma_{i\in s}(Y_i/\pi_i)$ as an estimator for Y based on an open survey. It is known in the literature (see Appendix 4 for a review) that under various super population modeling the model-expected variance of the Horvitz-Thompson (1952) estimator, namely $E_m V_p e^*(s,\underset{\sim}{Y})$, is less than $E_m V_p e(s,\underset{\sim}{Y})$ for any p-unbiased estimator $e(s,\underset{\sim}{Y})$ for $\underset{\sim}{Y}$ when both $e^*(s,\underset{\sim}{Y})$ and $e(s,\underset{\sim}{Y})$ are based on the same design or a common class of designs.

Also, $E_m V_{p_{(1)}} e^*(s, \underline{Y})$ is less than $E_m V_{p_{(2)}} e(s, \underline{Y})$ when $e^*(s, \underline{Y})$ is based on a design $p_{(1)}$ within a class of designs but $e(s, \underline{Y})$ is a p-unbiased estimator for Y but based on a design $p_{(2)}$ within a class of designs which may coincide with or may be wider than the class of designs containing $p_{(1)}$. That is, the class of strategies $(p_{(1)}, e^*(s, \underline{Y}))$ is optimal under a class of models against the competing class of strategies $(p_{(2)}, e(s, \underline{Y}))$. From (7.31) it follows that when this happens with an open survey, the corresponding results follow in favor of $(p, t^*(s, \underline{Z}))$ against $(p, e(s, \underline{Z}))$ with a fixed design p for both $t^*(s, \underline{Z})$ and $e(s, \underline{Z})$ or in favor of $(p_{(1)}, t^*(s, \underline{Z}))$ against $(p_{(2)}, e(s, \underline{Z}))$'s for a class of designs containing $p_{(1)}$ and the same or a wider class of designs containing $p_{(2)}$. The results noted above are summarized briefly in the following theorems and corollaries.

Theorem 7.7. If for any pR-unbiased estimator $e(s, \underline{Z})$ for Y, under any model,

$$E_m V_p e^*(s, \underline{Y}) \le E_m V_p [E_R e(s, \underline{Z})]$$

then $E_m V_{pR} t^*(s, \underline{Z}) \le E_m V_{pR} e(s, \underline{Z})$.

Corollary 1. If $E_m V_{p_{(1)}} e^*(s, \underline{Y}) \le E_m V_{p_{(2)}} e(s, \underline{Y})$ for every design $p_{(1)}$ in a class $\mathscr{P}_1$ of designs and for every design $p_{(2)}$ in a class $\mathscr{P}_2$ of designs containing $\mathscr{P}_1$, for every $e(s, \underline{Y})$ such that $E_{p_{(2)}} e(s, \underline{Y}) = Y$ for every $p_{(2)}$ in $\mathscr{P}_2$, then

$$E_m V_{p_{(1)}R} t^*(s, \underline{Z}) \le E_m V_{p_{(2)}R} e(s, \underline{Z})$$

for every $e(s, \underline{Z})$ such that $E_{p_{(2)}R} e(s, \underline{Z}) = Y$.

Corollary 2. $V_p e^*(s, \underline{Y}) \le V_p e(s, \underline{Y})$ for any p-unbiased estimator $e(s, \underline{Y})$ for Y implies that $V_{pR} t^*(s, \underline{Z}) \le V_{pR} t(s, \underline{Z})$, where $t(s, \underline{Z})$ is the derived estimator corresponding to $e(s, \underline{Y})$. Incidentally, it is easy to work out the variance formula:

$$\begin{aligned} V_{pR}(t^*(s, \underline{Z})) &= V_{pR}\left(\frac{1}{C}\sum \frac{Z_i^*}{\pi_i}\right) = \sum_i^N Y_i^2 \left(\frac{1}{C\pi_i} - 1\right) \\ &+ \sum_{i \neq j}^N \sum^N \left(\frac{\pi_{ij}}{\pi_i \pi_j} - 1\right) Y_i Y_j \\ &+ \frac{1-C}{C^2}\left(\sum_{9j} X_j^2\right) \sum_1^N \frac{1}{\pi_i} - \frac{(1-C)^2}{C^2} \mu_X^2 \sum_1^N \frac{1}{\pi_i} \\ &- \frac{2(1-C)}{C} \mu_X \sum_1^N \frac{Y_i}{\pi_i} \end{aligned}$$

A pR-unbiased estimator for this variance is

$$v = \frac{1}{C^2} \sum\sum_{i<j\in s} \frac{\pi_i \pi_j - \pi_{ij}}{\pi_{ij}} \left(\frac{Z_i^*}{\pi_i} - \frac{Z_j^*}{\pi_j} \right)^2 + \frac{1-C}{C^2} \sum_{i\in s} \frac{Z_i^2}{\pi_i}$$

$$+ \frac{1-C}{C} N \sum_1^M \mathcal{G}_j X_j^2 - \frac{(1-C)^2}{C^2} \mu_X^2 \sum_{i\in s} \frac{1}{\pi_i}$$

$$- \frac{2(1-C)}{C^2} \mu_X \sum_{i\in s} \frac{Z_i^*}{\pi_i}$$

Because of Theorem 7.7 and Corollary 1, many of the optimality results on strategies of sampling finite populations due, for example, to Godambe and Joshi (1965), Godambe and Thompson (1973, 1977), Cassel *et al.* (1976), and many others, obviously acquire extended counterparts for RR surveys. Comparisons between Horvitz-Thompson's and other strategies by Rao (1966a, b), Vijayan (1966), and Chaudhuri and Arnab (1979) also have obvious RR analogs.

Some details given by Chaudhuri and Adhikary (1981) and Chaudhuri (1986) in a slightly different version are recounted below presenting specific results. Variance expressions and pR-unbiased estimators for variances are given to facilitate practical applications of RR techniques.

In order that we may cover various modes of eliciting randomized responses, let us consider the unified setup as follows. From each sampled individual, labeled i, let an RR Z_i be obtained, independently of each other, so that

$$E_R(Z_i) = a_i Y_i + b_i \qquad E_R(Z_i^2) = a_i Y_i^2 + d_i$$

$$V_R(Z_i) = a_i(1-a_i) Y_i^2 - 2a_i b_i Y_i + (d_i - b_i^2)$$

with $0 < a_i < 1$, b_i, $d_i (\geq b_i^2)$, $i = 1, \ldots, N$, as certain known numbers. Considering $r_i = (Z_i - b_i)/a_i$, $i = 1, \ldots, N$, we will write $V_i = V_R(r_i) = \alpha_i Y_i^2 + \beta_i Y_i + \varepsilon_i$, where $\alpha_i = (1-a_i)/a_i$, $\beta_i = -2b_i/a_i$, and $\varepsilon_i = (d_i - b_i^2)/a_i^2$. The case of a single attribute is covered if y is an indicator variable. A special case of this is as we considered earlier with $a_i = C$, $b_i = (1-C)\mu_X$, $i = 1, \ldots, N$. This RR model originated from a suggestion due to Thompson (1984) expressed in her comments on the works of Adhikary *et al.* (1984).

For $r_i = (Z_i - b_i)/a_i$, $i = 1, \ldots, N$, let $\underline{r} = (r_1, \ldots, r_N)$ and $r = \Sigma_1^N r_i$. We will consider two classes of estimators for Y based on $(s, r_i | i \in s)$: (1) C_1 consisting of estimators $e = e(s, \underline{r})$ such that $E_p\, e(s, \underline{r}) = r$ and hence that $E_{pR} e(s, \underline{r}) = Y$, and (2) C_2 consisting of estimators $e_b = e_b(s, \underline{r}) = b_s + \Sigma_{i\in s} b_{si} r_i$ such that $E_p b_s = 0$, $\Sigma_{s \ni i} b_{si} p(s) = 1$ for every i so that $E_p e_b = r$ and hence that

$E_{pR}e_b = Y$. Of course C_2 is a subclass of C_1. In particular, we will write $e^* = e^*(\underline{r}) = \Sigma_{i \in s} r_i/\pi_i = e^*(s, \underline{r})$ to denote an RR version of the Horvitz-Thompson (1952) estimator. We will consider two superpopulation models M_1 and M_2 postulating distributions of $\underline{Y}$ treated as a random vector such that:

1. Under M_1, Y_i's are independent with

$$E_m(Y_i) = \theta_i \qquad V_m(Y_i) = \sigma_i^2$$

2. Under M_2, $E_m(Y_i) = \theta_i$, $V_m(Y_i) = \sigma_i^2$, and $C_m(Y_i, Y_j) = \rho\sigma_i\sigma_j$ with $-1/(N-1) < \rho < 1$, $\sigma_i > 0$ and ρ, σ_i, θ_i as otherwise unknown real numbers.

Of course, we assume that E_m, E_p, and E_R commute with one another. In case some positive numbers W_i's with $W = \Sigma W_i$ are known such that $\theta_i = \beta W_i$, $\sigma_i = \sigma W_i$, where β and σ are unknown and $0 < \beta, \sigma < \infty$, we denote corresponding M_1 and M_2 as $M_1(W)$ and $M_2(W)$, respectively.

The following results are worth consideration.

Theorem 7.8. Among estimators in C_1 based on any design p, there does not exist one with the uniformly minimum variance (UMV).

Proof. If $\bar{e} = \bar{e}(s, \underline{r})$ is a UMV estimator in C_1 for Y, for a fixed $\underline{Y}_0 = (Y_{10}, \ldots, Y_{i0}, \ldots, Y_{No})$ with $Y_0 = \Sigma_1^N Y_{i0}$ and a vector $\underline{r}_0 = (r_{10}, \ldots, r_{No})$ corresponding to $\underline{r}$ with $E_R(r_{i0}) = Y_{i0}$, $i = 1, \ldots, N$, and $r_0 = \Sigma_1^N r_{i0}$, let an estimator be taken as

$$e' = e'(s, \underline{r}) = \bar{e}(s, \underline{r}) - E_R\bar{e}(s, \underline{r}_0) + Y_0$$

Then $V_{pR}e'(s, \underline{r}_0) = E_p V_R \bar{e}(s, \underline{r}_0) < V_{pR}\bar{e}(s, \underline{r}_0)$ unless $E_R\bar{e}(s, \underline{r}_0) = E_p E_R \bar{e}(s, \underline{r}_0) = Y_0$, for every s, which is ruled out for any design not amounting to a census design. Hence the proof is complete.

Remark. There does not exist any estimator $e = e(s, \underline{r})$ in C_1 such that $V_{pR}(e) \le V_{pR}(e^*)$ uniformly in $\underline{r}$ and strictly so for at least one $\underline{r}$. That is e^* is admissible. It is left as an exercise for the reader to prove this.

Theorem 7.9A. For every estimator $e_b = e_b(s, \underline{r})$ in C_2, under model $M_2(W)$,

$$E_m E_{p_n(W)R}(e_b - Y)^2 \ge \frac{W}{n}\sum_1^N \frac{E_m V_R(r_i)}{W_i} + (1-\rho)\sigma^2\left(\frac{W^2}{n} - \sum_i^N W_i^2\right)$$

$$= E_m E_{p_n(W)R}(e^* - Y)^2$$

Proof. Let $e_b(s, \underline{Y}) = b_s + \sum_{i \in s} b_{si} Y_i$, $e^*(\underline{Y}) = \sum_{i \in s}(Y_i/\pi_i)$. Then

$$V_{pR}(e_b) = V_p e_b(s, \underline{Y}) + \sum_1^N V_R(r_i) \sum_{s \supset i} b_{si}^2 p(s)$$
$$V_{pR} e_b(s, \underline{r}) - V_{pR} e^*(\underline{r}) \geq V_p e_b(s, \underline{Y}) - V_p e^*(\underline{Y})$$

since by the Cauchy-Schwarz inequality, $\Sigma_{s \supset i} b_{si}^2 p(s) \geq 1/\pi_i$ for each i. Therefore, under M_2,

$$E_m[V_{pR} e_b(\underline{r}) - V_{pR} e^*(\underline{r})]$$
$$\geq \left\{ V_p\left(b_s + \sum_{i \in s} b_{si}\theta_i\right) + V_p\left(\sum_{i \in s} b_{si}\sigma_i\right) + (1-\rho)\sum_1^N \sigma_i^2 \left[\sum_{s \supset i} b_{si}^2 p(s) - 1\right]\right\}$$
$$-\left[V_p\left(\sum_{i \in s} \frac{\theta_i}{\pi_i}\right) + V_p\left(\sum_{i \in s} \frac{\sigma_i}{\pi_i}\right) + (1-\rho)\sum_1^N \sigma_i^2 \left(\frac{1}{\pi_1} - 1\right)\right]$$

Hence the result follows giving optimality of $(p_n(W), e^*(\underline{r}))$'s under $M_2(W)$, among $(p_n(W), e_b(\underline{r}))$'s. The estimator $e^*(\underline{r})$ was suggested by Thompson (1984).

Remark. A limitation of the optimality above is that all the competing estimators are based on p_n(W).

Theorem 7.9B. Under model $M_1(W)$, for every $e(\underline{r})$ in C_1,

$$E_m E_{p_n(\mathrm{W})\mathrm{R}}[e(\underline{r}) - Y]^2 \geq E_m E_{p_n(\mathrm{W})\mathrm{R}}[e^*(\underline{r}) - Y]^2$$

Proof. Let $e(s, \underline{Y}) = e^*(s, \underline{r}) + h(s, \underline{r})$, so that $h(s, \underline{r})$ is free of r_i's for $i \notin s$ and $E_p h(s, \underline{r}) = 0$, the latter implying that $\Sigma_{s \supset i} h(s, \underline{r}) p(s) = -\Sigma_{s \not\supset i} h(s, \underline{r}) p(s)$. Then

$$V_{pR} e(s, \underline{r}) = E_p V_R e^*(s, \underline{r}) + E_p V_R h(s, \underline{r}) + V_p E_R e(s, \underline{r})$$

since

$$E_p C_R(e^*(s, \underline{r}), h(s, \underline{r})) = E_R \sum_1^N \frac{r_i - Y_i}{\pi_i} \sum_{s \supset i} h(s, \underline{r}) p(s)$$
$$= -\sum_i E_R \frac{r_i - Y_i}{\pi_i} E_R \sum_{s \not\supset i} h(s, \underline{r}) p(s) = 0$$

Now

$$E_m[V_{pR} e(s, \underline{r}) - V_{pR} e^*(s, \underline{r})]$$
$$= E_m[E_p V_R h(s, \underline{r}) + V_p E_R e(s, \underline{r}) - V_p E_R e^*(s, \underline{r})]$$

Writing e for $e(s, \underline{r})$, e^* for $e^*(s, \underline{r})$, and h for $h(s, \underline{r})$, we have under M_1,

$$
\begin{aligned}
E_m V_p E_R(e) &= E_m E_p[E_R(e) - E_{pR}(e)]^2 = E_m E_p[E_R(e) - Y]^2 \\
&= E_p E_m[(E_R e - E_{mR} e) + (E_{mR} e - E_m Y) - (Y - E_m Y)]^2 \\
&= E_p V_m(E_R e) + E_p \Delta_m^2(E_R e) - V_m(Y)
\end{aligned}
$$

and hence

$$
\begin{aligned}
&E_m V_p(E_R e) - E_m V_p(E_R e^*) \\
&\quad = E_p V_m(E_R e) - E_p V_m(E_R e^*) + E_p \Delta_m^2(E_R e) - E_p \Delta_m^2(E_R e^*) \\
&\quad = E_p V_m(E_R h) + E_p \Delta_m^2(E_R e) - E_p \Delta_m^2(E_R e^*)
\end{aligned}
$$

on writing

$$
\begin{aligned}
\Delta_m(E_R e) &= E_{mR}(e) - E_m(Y) \\
\Delta_m(E_R e^*) &= E_{mR}(e^*) - E_m(Y)
\end{aligned}
$$

and using the fact that

$$
\begin{aligned}
&E_p C_m(E_R e^*, E_R h) \\
&\quad = E_m \sum_1^N \frac{Y_i - \theta_i}{\sigma_i} \sum_{s \supset i} E_R h(s, \underline{r}) = -\sum_1^N E_m \frac{Y_i - \theta_i}{\pi_i} E_m \sum_{s \not\supset i} E_R h(s, \underline{r}) = 0
\end{aligned}
$$

Therefore,

$$
\begin{aligned}
&E_m[V_{pR} e(s, \underline{r}) - V_{pR} e^*(s, \underline{r})] \\
&\quad = E_m E_p V_R h(s, \underline{r}) + E_p V_m(E_R h) + E_p \Delta_m^2(E_R e) - \Delta_m^2(E_R e^*)
\end{aligned}
$$

and the result follows on noting that under $M_1(\mathrm{W})$, $\Delta_m(E_R e^*) = 0$.

Remark. The same limitation applies as earlier.

Incidentally, note that for any fixed-sample-size design p, we have the following:

$$
\begin{aligned}
\text{(a)}\quad V = V_{pR} e^*(s, \underline{r}) &= V_p e^*(s, \underline{Y}) + E_p\left[\sum_{i \in s} \frac{V_R(r_i)}{\pi_i^2}\right] \\
&= \sum_{i<j}^N \sum^N (\pi_i \pi_j - \pi_i)\left(\frac{Y_i}{\pi_i} - \frac{Y_j}{\pi_j}\right)^2 + \sum_{i=1}^N \frac{V_R(r_i)}{\pi_i}
\end{aligned}
$$

and

$$
\text{(b)}\quad v = \sum_{i<} \sum_{j \in s} \frac{\pi_i \pi_j - \pi_{ij}}{\pi_{ij}} \left(\frac{r_i}{\pi_i} - \frac{r_j}{\pi_j}\right)^2 + \sum_{i \in s} \frac{a_i}{\pi_i}(\alpha_i r_i^2 + \beta_i r_i + \gamma_i)
$$

is a pR-unbiased estimator for V, recalling that

$$
\alpha_i = \frac{1}{a_i} - 1 \qquad \beta_i = \frac{2b_i}{a_i}, \qquad \gamma_i = \frac{d_i - b_i^2}{a_i^2}
$$

We will write $\hat{\sigma}_i^2 = a_i(\alpha_i r_i^2 + \beta_i r_i + \gamma_i)$ to get $E_R(\hat{\sigma}_i^2) = \sigma_i^2 = V_R(r_i)$.

Let M_3 be a model postulating the distribution of $\underline{Y}$ such that $E_m(Y_i) = \theta_i$, $V_m(Y_i) = \sigma_i^2$, $\theta = \Sigma\theta_i$. $\sigma = \Sigma\sigma_i$, $C_m(Y_i, Y_j) = 0$, $i \neq j$, with θ_i in $(-\infty, +\infty)$, σ_i^2 in $(0, \infty)$, $i = 1, \ldots, N$. Let $p_n(\theta)$ denote the class of fixed (n)-sample-size designs for which $\pi_i = n\theta_i/\theta$, $i = 1, \ldots, N$. Then the following holds.

Theorem 7.10. Under model M_3,

$$E_m V_{p_n R}[e_b(s, \underline{r})] \geq E_m V_{p_n(\theta) R} e^*(\underline{r})$$

Proof. $V_{pR} e_b(s, \underline{r}) = V_p(b_s + \Sigma_{i \in s} b_{si} Y_i) + \Sigma_i V_R(r_i) \Sigma_{s \supset i} b_{si}^2 p(s)$,

$$V_{pR} e^*(\underline{r}) = V_p\left(\sum_{i \in s} \frac{Y_i}{\pi_i}\right) + \sum_1^N \frac{V_R(r_i)}{\pi_i}$$

Hence

$$V_{pR} e_b(s, \underline{r}) - V_{pR} e^*(\underline{r}) \geq V_p\left(b_s + \sum_{i \in s} b_{si} Y_i\right) - V_p\left(\sum_{i \in s} \frac{Y_i}{\pi_i}\right)$$

But

$$E_m V_p\left(b_s + \sum_{i \in s} b_{si} Y_i\right) = E_p V_m\left(b_s + \sum_{i \in s} b_{si} Y_i\right)$$

$$+ E_p\left(b_s + \sum_{i \in s} b_{si}\theta_i - \theta\right)^2 - V_m(\mathrm{Y})$$

$$= \sum_1^N \sigma_i^2 \left[\sum_{s \supset i} b_{si}^2 p(s) - 1\right] + E_p\left(b_s + \sum_{i \in s} b_{si}\theta_i - \theta\right)^2$$

$$E_m V_p\left(\sum_{i \in s} \frac{Y_i}{\pi_i}\right) = \sum_1^N \sigma_i^2 \left(\frac{1}{\pi_i} - 1\right) + E_p\left(\sum_{i \in s} \frac{\theta_i}{\pi_i} - \theta\right)^2$$

The result follows immediately.

Remark. This is an extension of Godambe's (1955) open survey result. In this uncorrelated case competitors $(p_n, e_b(s, \underline{r}))$ with $(p_n(\theta), e^*(\underline{r}))$ are wider than allowed in Theorems 7.8 and 7.9A.

Remark

(a) For any e_b, $V_p e_b(s, \underline{Y}) - V_p e^*(s, \underline{Y}) \geq 0$ uniformly in $\underline{Y} \Rightarrow V_{pR} e_b(s, \underline{r}) - V_{pR} e^*(s, \underline{r}) \geq 0$ uniformly in $\underline{r}$.

(b) For any model, in fact, $E_m[V_p e_b(s, \underline{Y}) - V_p e^*(s, \underline{Y})] \geq 0$ $\Rightarrow E_m[V_{pR} e_b(s, \underline{r}) - V_{pR} e^*(s, \underline{r})] \geq 0$.

(c) Unlike Theorem 7. 7 and Corollaries 1 and 2 above (Adhikary *et al.* 1984), the foregoing approach here yields no corresponding result about $e^*(s,\underset{\sim}{r})$ versus $e(s,\underset{\sim}{r})$ comparable to (a) and (b).

Next consider the random permutation model M_4, which stipulates the following. Let $\underset{\sim}{A}=(A_1,\ldots,A_1,\ldots,A_N)$ be a vector of known positve numbers A_i's such that $\Sigma_i^N A_1=n$, $R_i=Y_i/A_i$, $i=1,\ldots,N$, $\underset{\sim}{R}=(R_1,\ldots,R_i,\ldots,R_N)$. Let for a permutation $\mathscr{P}=(P_1,\ldots,P_N)$ of $I=(1,\ldots,N)$, $\mathscr{P}\underset{\sim}{R}=(R_{P_1},\ldots,R_{P_N})$, and $\underset{\sim}{Y}=(Y_1,\ldots,Y_i,\ldots,Y_N)$ be a random vector so distributed that each of the $N!$ points $(R_{P_1},\ldots,R_{P_N})$ is assigned a probability $1/N!$ of appearing as a permutation of the fixed vector $\underset{\sim}{R}$. We will write $\Sigma\mathscr{P}$ to denote the sum over $N!$ terms corresponding to permutation of the coordinates of I_N. Under this model M_4, we have the following theorem, on writing $p_n(A)$ for any design of the class of designs p_n such that $\pi_i=A_i$, $i=1,\ldots,N$ (see Godambe and Thompson, 1973).

Theorem 7.11. Under model M_4:
(a) For every p of the form $p_n(A)$,

$$E_m E_R E_p(e^*(s,\underset{\sim}{r})-Y)^2$$

has a common value, say,

$$E_m E_R E_{p_n(A)}[e^*(s,\underset{\sim}{r})-Y]^2$$

(b) For any $e=e(s,\underset{\sim}{r})$ with $E_R(e)=\Sigma_1^N r_i$, one has

$$E_m E_R E_{p_n}[e(s,\underset{\sim}{r})-Y]^2 \geq E_m E_R E_{p_n(A)}[e^*(s,\underset{\sim}{r})-Y]^2$$

Proof. We will prove (b) first as follows. We may note that

$$\begin{aligned} Y &= \sum Y_i = \sum A_i R_i = E_{pR}(e) = E_p E_R e(s,\underset{\sim}{r}) = E_p h(s,\underset{\sim}{Y}) \\ &= \sum p(s) h(s,\underset{\sim}{Y}) = \sum p(s) K(s,\underset{\sim}{R}) \end{aligned}$$

say. Further, observe that

$$\begin{aligned} E_m E_R E_{p_n}[e(s,\underset{\sim}{r})-Y]^2 &= E_m\left[E_R \sum_s p(s) e^2(s,\underset{\sim}{r}) - Y^2\right] \\ &= E_m\left[\sum_s p(s) f^2(s,\underset{\sim}{Y}) - Y^2\right] \quad \text{(say)} \\ &= E_m\left[\sum_s p(s) g^2(s,\underset{\sim}{R}) - \left(\sum A_i R_i\right)^2\right] \quad \text{(say)} \\ &= \frac{1}{N!}\sum_{\mathscr{P}}\left[\sum_s p(s) g^2(s,\mathscr{P}\underset{\sim}{R}) - \left(\sum_1^N A_i R_{P_i}\right)^2\right] \end{aligned}$$

$$E_m E_{p_n \mathrm{R}} e(s, \underline{r}) = E_m \sum_1^{\mathrm{N}} A_i R_i = \frac{1}{N!} \sum_{\mathscr{P}} \left(\sum_1^{\mathrm{N}} A_i R_{\mathrm{P}_i} \right)$$

$$= \frac{(N-1)!}{N!} \left(\sum_1^{\mathrm{N}} A_i \right) \left(\sum_1^{\mathrm{N}} R_i \right) = \frac{n}{N} \sum R_i = n\bar{R} \quad \text{(say)}$$

Thus

$$n\bar{R} = E_m E_{\mathrm{R}} E_p e(s, \underline{r}) = E_m E_p h(s, \underline{Y})$$

$$= E_m \sum p(s) K(s, \underline{R}) = \frac{1}{N!} \sum_{\mathscr{P}} \sum_s p(s) K(s, \mathscr{P} \underline{R})$$

We may denote by $\mathscr{P}^p$ a design that stipulates choosing a sample s, say, $s = (j_1, \ldots, j_n)$ of elements out of I_{N} with probability $p(s)$ and observing R_i's for the subscripts obtained for a permutation $\mathscr{P}$ of $j_1, \ldots, j_n$. A typical unbiased estimator for $n\bar{R}$ based on such a design is $K = K(s, \mathscr{P} \underline{R})$, which is free of R_i's for $i \neq P_{j_1}, \ldots, P_{j_n}$ and has a variance

$$\frac{1}{N!} \sum_{\mathscr{P}} \sum_s p(s) K^2(s, \mathscr{P} \underline{R}) - (n\bar{R})^2$$

By $\bar{d}$ we denote the reduced data corresponding to sample observations d, suppressing the labels in the sample and retaining only the values R_i. Let Σ_1 be the sum over $\mathscr{P}$ with $s, \bar{d}$ fixed, Σ_2 the sum over samples with a common $\bar{d}$, and Σ_3 the sum over $\mathscr{P}$ with a fixed $\bar{d}$ for a particular s. Then

$$\xi(\bar{d}) = \frac{\Sigma_2 \Sigma_1 K(s, \mathscr{P} \underline{R}) p(s)}{\Sigma_2 \Sigma_1 p(s)}$$

is an "order statistic" corresponding to a d that is given. By completeness, it equals $\Sigma_{i \in s} R_{\mathrm{P}_i}$.

By Rao-Blackwellization, it follows that

$$\frac{1}{N!} \sum_{\mathscr{P}} \sum_s p(s) K^2(s, \mathscr{P} \underline{R}) \geq \frac{1}{N!} \sum_{\mathscr{P}} \sum_s p(s) \left(\sum_{i \in s} R_{\mathrm{P}_i} \right)^2$$

$$= \frac{1}{N!} \sum_{\mathscr{P}} \sum_s p'(s) \left(\sum_{i \in s} R_{\mathrm{P}_i} \right)^2 \qquad (7.32)$$

for any design p' of the class p_n and also of the class $p_n(\mathrm{A})$. It follows that

$$E_m E_p [E_{\mathrm{R}} e(s, \underline{r}) - n\bar{R}]^2 \geq E_m E_{p'} \left(\sum_{i \in s} R_{\mathrm{P}_i} - n\bar{R} \right)^2$$

$$= E_m E_{p_n(\mathrm{A})} \left(\sum_{i \in s} R_i - n\bar{R} \right)^2$$

on subtracting $(n\bar{R})^2$ from both sides of (7.32). Hence

$$E_m E_p[E_R e(s, \underline{r}) - n\bar{R}]^2 \geq E_m E_{p_n(A)}\left(E_R \sum_{i \in s} \frac{r_i}{\pi_i} - n\bar{R}\right)^2$$

$$= E_m E_{p_n(A)}[E_R e^*(s, \underline{r}) - n\bar{R}]^2 \qquad (7.33)$$

But

$$E_m E_p E_R[e(s, \underline{r}) - Y]^2 = E_m E_p E_R\{[e(s, \underline{r}) - E_R e(s, \underline{r})] + [E_R e(s, \underline{r}) - E_m E_p E_R e(s, \underline{r})] - (Y - E_m Y)\}^2$$

$$= E_m E_p V_R[e(s, \underline{r})] + E_m E_p\{E_R[e(s, \underline{r}) - n\bar{R}]\}^2 - E_m(Y - E_m Y)^2$$

$$E_m E_p E_R[e^*(s, \underline{r}) - Y]^2 = E_m E_p V_R[e^*(s, \underline{r})] + E_m E_p[E_R e^*(s, \underline{r}) - n\bar{R}]^2 - E_m(Y - E_m Y)^2$$

Therefore, by (7.33),

$$E_m E_p E_R[e(s, \underline{r}) - Y]^2 - E_m E_p E_R[e^*(s, \underline{r}) - Y]^2 \geq E_m E_p V_R[e(s, \underline{r})] - E_m E_p V_R[e^*(s, \underline{r})] \qquad (7.34)$$

Also,

$$E_p V_R[e(s, \underline{r})] = E_p V_R[e^*(s, \underline{r}) + a(s, \underline{r})]$$

say, where

$$a(s, \underline{r}) = e(s, \underline{r}) - e^*(s, \underline{r})$$

so that $E_p a(s, \underline{r}) = \Sigma_s p(s) a(s, \underline{r}) = 0$. This implies that

$$\sum_{s \supset i} p(s) a(s, \underline{r}) = -\sum_{s \not\supset i} p(s) a(s, \underline{r}) \qquad (7.35)$$

Consequently,

$$E_p V_R[e(s, \underline{r})] = E_p\{V_R[e^*(s, \underline{r})] + V_R a(s, \underline{r}) + 2C_R[e^*(s, \underline{r}), a(s, \underline{r})]\}$$

$$\geq E_p V_R e^*(s, \underline{r}) \qquad (7.36)$$

The above holds since

$$E_p C_R[e^*(s, \underline{r}), a(s, \underline{r})]$$

$$= \sum_s p(s) E_R\left[\left(\sum_{i \in s} \frac{r_i - Y_i}{\pi_i}\right) a(s, \underline{r})\right]$$

$$= E_R\left[\sum_1^N \frac{r_i - Y_i}{\pi_i} \sum_{s \supset i} p(s)\, a(s, \underline{r})\right]$$

$$= -E_R\left[\sum_1^N \frac{r_i - Y_i}{\pi_i} \sum_{s \not\supset i} p(s)\, a(s, \underline{r})\right] = 0$$

by (7.35) and the independence of r_i's. Combining (7.34) and (7.36), one gets

$$E_m E_{p_n} E_R[e(s, \underline{r}) - Y]^2 \geq E_m E_{p_n(A)} E_R[e^*(s, \underline{r}) - Y]^2$$

This completes the proof of part (b) of Theorem 7.11. Proof of part (a) is also evident from the above.

7.5. MODIFICATIONS OF CERTAIN POPULAR SAMPLING STRATEGIES IN OPEN SURVEYS WHEN RESPONSES ARE RANDOMIZED

Optimality of sampling strategies is well known to hold only under exceptionally restrictive circumstances in case of open surveys. The same applies to randomized surveys as well, as is evident from the above. In open surveys, the usual practice is to contemplate restricting the choice among a few particular popular strategies only and to compare among them in terms of a general class of models in making an appropriate choice. For example, one often postulates a model, say M_5, stipulating the distribution of $\underline{Y}$ such that

$$E_m(Y_i) = \beta W_i \qquad \left.\begin{aligned} C_m(Y_i, Y_j) &= \sigma^2 W_i^g \qquad && \text{if } i = j \\ &= 0 && \text{if } i \neq j \end{aligned}\right\}$$

such that $W_i > 0, i = 1, \ldots, N$, with $\Sigma W_i = W$ as known numbers, β and σ as unknown positive numbers, and g an unknown number in the closed interval $[0, 2]$. In the randomized survey setup also, let us assume this model to hold and consider the modified versions of some well-known strategies.

As before, let the RRs be r_i, $i = 1, 2, \ldots$, with

$$E_R(r_i) = Y_i$$

$$V_R(r_i) = \left(\frac{1}{a_i} - 1\right) Y_i^2 - 2\frac{b_i}{a_i} Y_i + \frac{d_i - b_i^2}{a_i^2} = \sigma_i^2 \qquad \text{(say)} \qquad (i = 1, \ldots, N)$$

Consider the following RR strategies:

1. The design is $p_n(W)$, that is, a p_n design with $\pi_i = nW_i/W$, $i = 1, \ldots, N$, and the estimator is $e_1 = \Sigma_{i \in s} r_i/\pi_i$.
2. It is supposed that $N/n = k$ is an integer and the Rao-Hartley-Cochran (1962) sampling strategy is used with p_i's as known normed, positive size measures and P_i's as sums of p_i's falling in the ith group ($i = 1, \ldots, n$) while forming n random groups in implementing sample selection. The modified estimator is $e_2 = \Sigma_i (r_i/p_i) P_i$.

3. The Lahiri-Midzuno-Sen (1951, 1952, 1953) sampling scheme is used with p_i's as in strategy 2 above, and the estimator is $e_3 = \Sigma_{i \in s} r_i / \Sigma_{i \in s} p_i$ for a sample (of size n each) s chosen with probability

$$p(s) = \frac{1}{\binom{N-1}{n-1}} \sum_{i \in s} p_i$$

4. The Hansen-Hurwitz (1943) sampling scheme is used in n draws with p_i's as above and the estimator is $e_4 = \Sigma_{i \in s} (r_i / p_i) f_{is}$, where f_{is} is a frequency of i in sample s.

5. Probabilities are proportional to size without replacement sampling in $n=2$ draws with p_i as before, together with Murthy's (1957) symmerized Des Raj (1956) estimator modified as

$$e_5 = \frac{1}{2 - p_i - p_j} \left[\frac{r_i}{p_i}(1 - p_j) + \frac{r_j}{p_j}(1 - p_i) \right]$$

The corresponding original estimators may be denoted, respectively, as e_i' $(i = 1, \ldots, 5)$, where r_i is replaced by Y_i. It follows immediately that

$$E_1 = E_m V_{pR}(e_1) = E_m \left[V_p(e_1') + \sum_1^N \frac{\sigma_i^2}{\pi_i} \right]$$

$$E_2 = E_m V_{pR}(e_2) = E_m \left[V_p(e_2') + \sum_1^N \sigma_i^2 \frac{P_i}{p_i} \right]$$

$$E_3 = E_m V_{pR}(e_3) = E_m \left[V_p(e_3') + \frac{1}{\binom{N-1}{n-1}} \sum_s \frac{\sum_{i \in s} \sigma_i^2}{\sum_{i \in s} p_i} \right]$$

$$E_4 = E_m V_{pR}(e_4) = E_m \left[V_p(e_4') + \frac{1}{n} \sum_1^N \frac{\sigma_i^2}{p_i} \right]$$

$$E_5 = E_m V_{pR}(e_5) = E_m \left[V_p(e_5') + \sum_{i \neq j}\sum \frac{\sigma_i^2 p_j (1 - p_j)}{p_i (1 - p_i)(2 - p_i - p_j)} \right]$$

One can easily draw upon the works of Rao (1966a, b), Rao (1967), Hanurav (1967), Chaudhuri and Arnab (1979), and many others to compare the relative magnitudes of $E_m V_p(e_i')$, $i = 1, \ldots, 5$. Also, one may note that

$$E_m(\sigma_i^2) = \left(\frac{1}{a_i} - 1 \right)(\sigma^2 W_i + \beta^2 W_i^2) - 2\frac{b_i}{a_i}\beta W_i + \frac{d_i - b_i^2}{a_i^2} \qquad (i = 1, \ldots, N)$$

and hence proceed to compare among E_i's, $i=1, \ldots, 5$, to make an appropriate choice of a sampling strategy.

Since we may write

$$V_i = V_{pR}(e_i) = E_R[V_p(e_i)] + V_R[E_p(e_i)]$$

we obtain the following pR-unbiased estimators for them, respectively, as v_i, $i=1, \ldots, 5$, recalling that $\hat{\sigma}_i^2 = a_i(\alpha_i r_i + \beta_i r_i + e_i)$ satisfies $E_R(\hat{\sigma}_i^2) = \sigma_i^2$:

$$v_1 = \sum\sum_{i<j\in s} \frac{\pi_i\pi_j - \pi_{ij}}{\pi_{ij}} \left(\frac{r_i}{\pi_i} - \frac{r_j}{\pi_j} \right)^2 + \sum_{i\in s} \frac{\hat{\sigma}_i^2}{\pi_i}$$

$$v_2 = \frac{N-n}{n-1} \frac{1}{N} \sum_{i=1}^{n} P_i(r_i - e_2)^2 + \sum_{i=1}^{n} \hat{\sigma}_i^2 \frac{P_i}{p_i}$$

$$v_3 = -\sum\sum_{i<j\in s} d_{ij}(s) p_i p_j \left(\frac{r_i}{p_i} - \frac{r_j}{p_j} \right)^2 + \frac{\sum_{i\in s} \hat{\sigma}_i^2}{\sum_{i\in s} p_i}$$

where $d_{ij}(s)$ is to be chosen so that $d_{ij}(s)=0$ if $s \not\supset i, j$ but $\Sigma_{s\supset i,j}\, d_{ij}(s)p(s) = d_{ij}$ with

$$d_{ij} = \sum_{s\supset i,j} \frac{p(s|i)p(s|j)}{p(s)} - 1$$

and $p(s|k)$ is the conditional probability of choosing s given that k is the label chosen on the first draw,

$$v_4 = \frac{1}{n(n-1)} \sum_{i\in s} (r_i - e_4)^2 f_{is} + \frac{1}{n} \sum_{i\in s} \frac{\hat{\sigma}_i^2}{p_i} f_{is}$$

and

$$v_5 = \frac{(1-p_i)(1-p_j)(1-p_i-p_j)}{(2-p_i-p_j)^2} \left(\frac{r_i}{p_i} - \frac{r_j}{p_j} \right)^2 + \frac{1}{2-p_i-p_j} \left[\frac{\hat{\sigma}_i^2}{p_i}(1-p_j) + \frac{\hat{\sigma}_j^2}{p_j}(1-p_i) \right]$$

These estimators are suggested through Rao's (1979) works relating to p-unbiased estimators for $V_p(e_i')$'s.

REFERENCES

Adhikary, A. K., Chaudhuri, A., and Vijayan, K. (1984). Optimum sampling strategies for RR trials. *Internat. Statist. Rev.* **52**, 115–125.

Bellhouse, D. R. (1980). Linear models for RR designs. *J. Amer. Statist. Assoc.* **75**, 1001–1004.

Cassel, C. M. Särndal, C. E., and Wretman, J. H. (1976). Some results on generalized difference estimation and generalized regression estimation for finite population. *Biometrika* **63**, 615–620.

Chaudhuri, A. (1986). RR surveys of finite populations: a unified approach with quantitative data. To appear in *J. Statist. Plann. Inference.*

Chaudhuri, A., and Adhikary, A. K. (1981). On sampling strategies with RR trials and their properties and relative efficiencies. *Tech. Rep. ASC*/81/5, Indian Statistical Institute, Calcutta.

Chaudhuri, A., and Arnab, R. (1979). On the relative efficiencies of sampling strategies under a super population model. *Sankhyā Ser. C.* **41**, 40–43.

Eriksson, S. (1973a). Randomized interviews for sensitive questions. Ph.D. thesis, University of Gothenburg.

Eriksson, S. (1973b). A new model for RR. *Internat. Statist. Rev.* **41**, 101–113.

Godambe, V. P. (1955). A unified theory of sampling from finite populations. *J. Roy. Statist. Soc. Ser. B.* **17**, 269–278.

Godambe, V. P. (1965). A review of the contributions towards a unified theory of sampling from finite populations. *Internat. Statist. Rev.* **33**, 242–258.

Godambe, V. P. (1980). Estimation in RR trials. *Internat. Statist. Rev.* **48**, 29–32.

Godambe, V. P. and Joshi, V. M. (1965). Admissibility and Bayes estimation in sampling from finite populations. *Ann. Math. Statist.* **36**, 1707–1722.

Godambe, V. P., and Thompson, M. E. (1973). Estimation in sampling theory with exchangeable prior distributions. *Ann. Statist.* **1**, 1212–1221.

Godambe, V. P., and Thompson, M. E. (1977). Robust near optimal estimation in survey practice. *Bull. Internat. Statist. Inst.* **47**, 129–146.

Greenberg, B. G., Abul-Ela, Abdel-Latif, A., Simmons, W. R., and Horvitz, D. G. (1969). The unrelated question RR model-theoretical framework. *J. Amer. Statist. Assoc.* **64**, 520–539.

Hansen, M. H., and Hurwitz, W. N. (1943). On the theory of sampling from finite populations. *Ann. Math. Statist.* **14**, 333–362.

Hanurav, T. V. (1966). Some aspects of unified sampling theory. *Sankhyā Ser. A.* **28**, 175–204.

Hanurav, T. V. (1967). Optimum utilization of auxiliary information: πPS sampling of two units from a stratum. *J. Roy. Statist. Soc. Ser. B.* **29**, 374–391, **31**, 192–194 (1969) with a correction note.

Hege, V. S. (1965). Sampling designs which admit uniformly minimum variance unbiased estimators. *Bull. Calcutta Statist. Assoc.* **14**, 160–162.

Horvitz, D. G., and Thompson, D. J. (1952). A generalization of sampling without replacement from finite universes. *J. Amer. Statist. Assoc.* **47**, 663–685.

Lahiri, D. B. (1951). A method of sample selection providing unbiased ratio estimates. *Bull. Internat. Statist. Inst.* **33**, 133–140.

Lanke, J. (1975). Some contributions to the theory of survey sampling. Ph.D. thesis, University of Lund.

Midzuno, H. (1952). On the sampling system with probability proportional to sum of sizes. *Ann. Inst. Statist. Math.* **3**, 99–107.

Murthy, M. N. (1957). Ordered and unordered estimators in sampling without replacement. *Sankhyā* **18**, 379–390.

Pollock, K. H., and Bek, Y. (1976). A comparison of three RR models for quantitative data. *J. Amer. Statist. Assoc.* **71**, 884–886.

Raj, Des (1956). Some estimators in sampling with varying probabilities without replacement. *J. Amer. Statist. Assoc.* **51**, 269–284.

Rao, C. R. (1973). *Linear Statistical Inference and Its Applications*, 2nd ed. John Wiley, New York.

Rao, J. N. K. (1966a). Alternative estimators in PPS sampling for multiple characters. *Sankhyā Ser. A* **28**, 47–60.

Rao, J. N. K. (1966b). On the relative efficiency of some estimators in PPS sampling for multiple characteristics. *Sankhyā Ser. A* **28**, 61–70.

Rao, J. N. K. (1979). On deriving mean square errors and their non-negative unbiased estimators in finite population sampling. *J. Indian Statist. Assoc.* **17**, 125–136.

Rao, J. N. K., and Bellhouse, D. R. (1978). Optimal estimation of a finite population mean under generalized random permutation model. *J. Statist. Plann. Inference* **2**, 125–141.

Rao, J. N. K., Hartley, H. O., and Cochran, W. G. (1962). On a simple procedure of unequal probability sampling without replacement. *J. Roy. Statist. Soc. Ser. B* **24**, 482–491.

Rao, T. J. (1967). On the choice of a strategy for the ratio method of estimation. *J. Roy. Statist. Soc. Ser. B* **29**, 392–397.

Sen, A. R. (1953). On the estimate of the variance in sampling with varying probabilities. *J. Indian Soc. Agri. Statist.* **5**, 119–127.

Thompson, M. E. (1984). Discussion of paper by A. K. Adhikary, A. Chaudhuri and K. Vijayan (1984).

Vijayan, K. (1966). On Horvitz-Thompson and Des Raj estimation. *Sankhyā Ser. A* **28**, 87–92.

8

Application of RRT and Concluding Remarks

Reduced response rate and inflated response bias are the two evils usually linked with direct surveys relating to sensitive items of inquiry. Gambling habits, addiction to drugs and other intoxicants, alcoholism, proneness to tax evasion, induced abortions, drunken driving, history of past involvement in crimes, and homosexuality are some of the characteristics about which people usually do not wish to divulge information. An investigator of such issues of social importance is reluctant to ask the sampled subjects direct questions with respect to such sensitive, stigmatizing, and perhaps incriminating matters. In addition to these types of categorical variables, certain quantitative ones also often engage the attention of a social researcher, such as the accumulated wealth of an important business person or a superstar in the world of film or sports, money earned through clandestine or unfair means, money spent for illegal or immoral purposes, the extent of taxes and duties evaded, and so on. Naturally, the people involved are inclined to keep such matters confidential and are reluctant to divulge the truth. The technique of RR is one of the available devices aimed at cutting down the levels of nonresponse, willful misstatement and lies and thus trying to improve the efficacy of inferences and to eliminate bias in estimation. Of course, once devised, an RRT needs to be applied in an effective and convincing manner so that not only is it desirable that one trust that prima facie a specific RRT should appear to protect a respondent's privacy, but more important, that the latter will believe it to be so, and on his or her own volition, answer the query honestly. Given a proposed RRT, it is most difficult to make an appraisal about its efficacy, and once applied, it is quite hazardous to guess and measure its fruitfulness. Of course, theoretical exercises are not difficult to implement in measuring an RRT's potentiality, but it is not easy to gauge how well it may fare in practice. Nevertheless, RR techniques galore have been proposed in the literature we are reviewing here, and more are appearing with amazing rapidity. Many of the techniques we mentioned earlier have been put to use. Reference to the literature shows that use of some of these techniques resulted in success, whereas use of others did not. Still, it may not be unwise to guess that converse experiences may result in the use of these techniques in the future. We will recount from the literature some of the noteworthy applications and investigations into the efficacy of RRTs in practice.

Following Warner's (1965) reported application of his technique in his pioneering work, Abernathy *et al.* (1970), Horvitz *et al.* (1975, 1976), Liu *et al.* (1975), and most of the other authors cited earlier in the book illustrated applications of various RR devices for specific problems and examined their effectiveness numerically through case studies. They also made comparative studies among various alternative procedures. Their studies included such topics as alcoholism, drug addiction, tax evasion, and incidence of induced abortion. Some outstanding empirical investigations into the uses of RRTs are those of Brewer (1981), Shimizu and Bonham (1978), Krotki and Fox (1974), Krotki and McDaniel (1975), Locander *et al.* (1976), Goodstadt and Gruson (1975), and Mukhopadhyay (1980), among others. Summaries of some of these investigations have been presented by Moors (1985) and Emrich (1983). With the object of ascertaining the extent of marijuana consumption among Australians, Brewer (1981) compared traditional open survey methods with a few RR devices. He also reported on the work of Goodstadt *et al.* (1978), Wiseman *et al.* (1975–1976), and Zdep and Rhodes (1976) to point out that Australian experiences did not corroborate the thesis that RR has a conclusive advantage over direct queries with respect to response rate and response bias. This makes Brewer (1981) rather sceptical about the promise of the RR technique in general in delivering any real advantage over the usual survey methods. On the other hand, studies on induced abortions by Krotki and Fox (1974), Krotki and McDaniel (1975) and Shimizu and Bonham (1978), on drug consumption as carried out by Goodstadt and Gruson (1975), and more important, on bankruptcy and drunken driving by Locander *et al.* (1976) reveal that use of the RR technique resulted in reduced refusal rates and less manipulation of responses. Krotki and McDaniel (1975) estimated the number of induced abortions in Alberta in 1973 as 12,320, 3060, and 1150, respectively using RR, mail questionnaire returns and face-to-face interviews, confirming the superiority of RRT (vide Moors, 1985). In a study carried out in Chicago, Locander *et al.* (1976) compared four interview techniques: face-to-face interview, telephone interview, self-administered questionnaire, and unrelated-question RRT. They considered four items of increasing level of sensitivity: possession of a public library card, voting behavior, involvement in bankruptcy, and accusation of drunken driving. On these issues accurate data could be gathered from official records. Hence a comparison could be made of how the estimates for the four techniques deviated from the truths. From Locander *et al.*'s (1976) investigation and their review by Emrich (1983), one may observe that RRT is least effective in reducing overreporting of socially desirable acts and most effective in cutting down underreporting of socially undesirable acts. Of course, no method is decidedly superior to all the others in providing good estimators with respect to all types of "threatening" questions. A study by Chi *et al.* (1972) was also inconclusive as to the quality of RR vis-à-vis DR in estimating the incidence of abortions. Goodstadt and Gruson (1975) demonstrated the superiority of RRT in yielding reliable estimates for proportions of students consuming drugs. Brown and Harding (1973) compared an unrelated-question RRT to an anonymously administered 62-item questionnaire in a study concerning nontherapeutic use of drugs in the military. Their investigation vindicated the efficacy of RRT in certain cases. On the other hand, in a study on the use of illicit drugs in the army implemented by Brown (1975), the conventional DR technique was found to yield a higher response rate than RR although the estimates produced by the two methods did not differ greatly. But Zdep and Rhodes (1976), in an inquiry on child abusers, demonstrated that the RR device used by Folsom *et al.* (1973) secured a greater response rate than that obtained using two conventional self-administrated techniques. Besides these studies on qualitative and categorical characteristics, reports are also available regarding studies on quantitative variables. One noteworthy reference in this connection is that of Pollock and Bek (1976), who applied three RR devices for the purpose as given by Greenberg *et al.* (1969), Warner (1971), and Poole (1974). Albers (1982) (see Moors, 1985) gives additional details, also taking into account the question of protection of privacy.

Two disadvantages associated with RR techniques are mentioned by Emrich. First, it is often difficult to implement them, as they require special skills on the part of investigators and more time and trouble in explaining to the subject how RR works. Second, the respondents may not be inclined to give truthful replies when they see that an RR scheme has in-built provisions to accommodate contaminated responses that permit the respondents to lie. Nevertheless, taking into consideration the rapid growth in the literature, with continual innovation of existing RR devices and novel techniques and application appearing with consistent regularity, one may only conclude that popular interest in RR theory and practice is growing. Campbell and Joiner (1973), who demonstrated the superiority of RR over DR in yielding a higher response rate in studying the habits of regular marijuana smokers among students, recommend the introduction of RR theory and methods into regular undergraudate course in the social sciences, including statistics.

In conclusion, we refer the reader to Chapter 9 of this book, where we give an account of two recent case studies presenting empirical findings by the Indian Statistical Institute on two uses of RR devices, covering categorical and quantitative queries.

REFERENCES

Abernathy, J. R., Greenberg, B. G., and Horvitz, D. G. (1970). Estimates of induced abortion in urban North Carolina, *Demography* **7**, 19–29.

Albers, W. (1982). Simple RR procedures with bounded respondent risk for quantitative data. *Kwantitatieve Methoden* **8**, 35–46.

Brewer, K. R. W. (1981). Estimating marijuana usage using RR: Some paradoxical findings. *Austral. J. Statist.* **23**, 139–148.

Brown, G. H. (1975). Randomized inquiry vs. conventional questionnaire method in estimating drug usage rates through mail surveys. *Tech. Rep.* 75–14, Human Resources Research Organization.

Brown, G. H., and Harding, F. D. (1973). A comparison of methods of studying illicit drug usage. *Tech. Rep.* 73–9. Human Resources Research Organization.

Campbell, C., and Joiner, B. L. (1973). How to get answers without being sure your're asked the question? *Amer. Statist.* **27**, 229–232.

Chi, I. C., Chow, L. P., and Rider, R. V. (1972). The RR technique as used in the Taiwan: outcome of pregnancy study. *Stud. Fam. Plann.* **3**, 265–269.

Emrich, L. (1983). RR techniques. In: *Incomplete Data in Sample Surveys*, Vol 2, ed. W. G. Madow, I. Olkin, and D. B. Rubin. Academic Press, New York, pp. 73–80.

Folsom, R. E., Greenberg, B. G., Horvitz, D. G., and Abernathy, J. R. (1973). The two alternate questions RR model for human surveys. *J. Amer. Statist. Assoc.* **68**, 525–530.

Goodstadt, M. S., and Gruson, V. (1975). The RR technique: a test on drug use. *J. Amer. Statist. Assoc.* **70**, 814–818.

Goodstadt, M. S., Cook, G., and Gruson, V. (1978). The validity of reported drug use: the RR technique. *Internat. J. Addiction* **13**, 359–367.

Greenberg, B. G., Abul-Ela, Abdel-Latif, A., Simmons, W. R., and Horvitz, D. G. (1969). The unrelated question RR model—theoretical framework. *J. Amer. Statist. Assoc.* **64**, 520–539.

Horvitz, D. G., Greenberg, B. G., and Abernathy, J. R. (1975). Recent developments in RR designs. In: *A Survey of Statistical Designs and Linear Models*, ed. J. N. Srivastava. North Holland, Amsterdam, pp. 271–285.

Horvitz, D. G., Greenberg, B. G., and Abernathy, J. R. (1976). RR: a data gathering device for sensitive questions. *Internat. Statist. Rev.* **44**, 181–196.

Krotki, K. J., and Fox, B. (1974). The RR technique, the interview and self-administered questionnaire: an empirical comparison of fertility reports. *Proc. ASA Soc. Sci. Sec.*, 367–371.

Krotki, K. J., and McDaniel, S. A. (1975). Three estimates of illegal abortions in Alberta, Canada: survey, mailback questionnaire and RR technique, in *Proceedings of the* 40*th Session, International Statistical Institute* 46/4, 67–69.

Liu, P. T., Chow, L. P., and Mosley, W. H. (1975). Use of the RR technique with a new randomizing device. *J. Amer. Statist. Assoc.* **70**, 329–333.

Locander, W., Sudman, S., and Bradburn, N. (1976). An investigation of interview method, threat and response distortion. *Proc. ASA Soc. Statist. Sec.*, 21–27.

Moors, J. J. A. (1985). Estimation in truncated parameter spaces. Ph.D. thesis, Katholieke Hogeschool, Tilburg.

Mukhopadhyay, P. (1980). A survey on the socio-economic conditions of some college students of Calcutta. Project report, Indian Statistical Institute, Calcutta.

Pollock, K. H., and Bek, Y. (1976). A comparison of three RR models for quantitative data. *J. Amer. Statist. Assoc.* **71**, 884–886.

Poole, W. K. (1974). Estimation of the distribution function of a continuous type random variable through RR. *J. Amer. Statist. Assoc.* **69**, 1002–1005.

Shimizu, I. M., and Bonham, G. S. (1978). RR technique in a national survey. *J. Amer. Statist. Assoc.* **73**, 35–39.

Warner, S. L. (1965). RR: a survey technique for eliminating evasive answer bias. *J. Amer. Statist. Assoc.* **60**, 63–69.
Warner, S. L. (1971). The linear RR model. *J. Amer. Statist. Assoc.* **66**, 884–888.
Wiseman, F., Moriarty, M., and Schafer, M. (1975–1976). Estimating public opinion with the RR model. *Public Opinion Quarterly* **39**, 507–513.
Zdep, S. M., and Rhodes, I. N. (1976) Making the RR technique work. *Public Opinion Quarterly* **40**, 531–537.

9

Case Studies

9.1. A SURVEY OF THE SOCIOECONOMIC CONDITIONS OF COLLEGE STUDENTS IN CALCUTTA WITH EMPHASIS ON DRUG HABITS

A project to carry out this survey was undertaken at the Indian Statistical Institute, Calcutta, in August-September 1980 by a team headed by P. Mukhopadhyay (1980), with technical guidance from A. Chaudhuri and A. C. Mukhopadhyay. The object was to estimate proportions of drug addicts among medical and nonmedical students in Calcutta at the undergraduate level and to study their variations over certain socio-economic groupings.

For this survey "drug" was taken to mean any substance that, taken into a living organism, may modify one or more of its functions. "Drug abuse" was understood as self-administration of a drug for nonmedical reasons, in quantities and frequencies that might impair a person's ability to function effectively and which may result in social, physical, or emotional harm. "Drug dependence" was taken to mean a state, psychic and sometimes also physical, resulting from the interaction between a living organism and a drug, characterized by behavioral and other responses, including a compulsion to take the drug on a continuous or periodic basis in order to experience its psychic effects and sometimes to avoid the discomfort of its absence. Three categories of drugs were distinguished: (1) sleeping pills, (2) alcohol and (3) other drugs. The reference period was taken as 1 week preceding the date of the survey for categories (1) and (2) and 1 month for (3). In addition, questions were asked about "smoking" habits among women students, as smoking is not common among women in India.

A suitably stratified sample was taken and a spinner was used to obtain an RR on each item independently, with a probability of $P=0.70$ referring to the stigmatizing character in each case. The findings are summarized in Table 9.1.

The nonresponse rate was negligible; in fact, many of the medical students volunteered to give direct responses. But in order to obtain unbiased estimates, willing students were also requested to utilize the RR technique. As the optional randomized response technique was not quite developed at that time, the available direct responses were not properly combined with the randomized responses in getting unbiased estimators. Instead, the investigators themselves applied randomization on behalf of those who responded directly.

9.2. RANDOMIZED RESPONSE SURVEY WITH SENSITIVE QUANTITATIVE CHARACTERS: A CASE STUDY

Arun Kumar Adhikary (1982), with technical guidance from A. Chaudhuri, in July 1982 carried out a survey among a group of workers at the Indian Statistical Institute to estimate their (1) total earnings from work in excess of that covered by their regular salaries, and (2) total expenses on drinking, gambling, and so on, over a period of 6 months just prior to the date of the survey. On the basis of their gross salaries for June 1982, the workers were classified into five strata.

The randomizing device was used in the following manner. A respondent was asked to note the color of a bead at the window of a box, containing 44 red and 36 black beads, when thoroughly shaken up. If a red bead was seen, the respondent would report the true variable value less a constant supplied to him or her. If a black bead was seen, the respondent would instead report the value of a ticket chosen randomly from another box [actually, two values from two boxes separately for items (1) and (2) as above] containing a given number of tickets with certain numbers inscribed on them (with the numbers 1, 2, 3, . . . inscribed on the reverse side, for random selection). Those values were chosen separately for the various strata, anticipating the ranges of possible true values. The numbers on the tickets were summed to zero. This is why a constant had to be deducted from the true values, to secure parity among the true and anticipated values,

TABLE 9.1. Percentage distribution of drug users and smokers among certain categories of Calcutta colleges, 1980[a]

	Male students				Female students	
	Medical		Nonmedical			
	Hostelliers	Day Scholars	Hostelliers	Day Scholars	Medical Hostelliers	Day Scholars
1. Smokers (women only)	—	—	—	—	10.0	5.9
2. Users of sleeping pills	12.5	7.2*	4.7	4.8	20.0	4.9
	(261)	(79)	(90)	(89)	(50)	(97)
3. Alcohol consumers	25.0	6.1*	4.8	8.0	11.5	4.7
	(261)	(79)	(90)	(89)	(50)	(97)
4. Users of other drugs	22.5	30.0	5.1*	20.5	13.9	4.9
	(261)	(79)	(90)	(89)	(50)	(97)

[a] Figures in parentheses indicate the sample sizes on which the estimates were based. Warner's estimates were used except for the starred entries, for which Warner's estimates were negative. For these, Bayes' estimates were used, following the work of Mukhopadhyay and Halder (1980).

in case it was to be reported. Of course, separate constants were used for the two items and for the various strata.

A 50% sample was taken, separately from each stratum, employing an IPPS (inclusion probability proportional to size) scheme following Sengupta's (1981) method and taking net take-home salaries as size measures. For estimation, a version of the Horvitz-Thompson estimator applicable to RR situations, as discussed in Section 7.4, was employed.

To present certain details, let us denote by $y^{(1)}$ and $y^{(2)}$, respectively, the two sensitive characters of interest: earnings through extra work and expenses on drinking and gambling. The hth stratum ($h=1, \ldots, 5$) consists of N_h identifiable units, yielding the vectors

$$\underline{Y}_h^{(j)}=(Y_{h1}^{(j)}, \ldots, Y_{h\mathrm{N}_h}^{(j)}), \qquad (j=1, 2)$$

of values. The problem is to estimate the stratum totals

$$Y_h^{(j)}=\sum_1^{\mathrm{N}_h} Y_{hi}^{(j)}, \qquad (j=1, 2)$$

and then the population totals

$$Y^{(j)}=\sum_h Y_h^{(j)}, \qquad (j=1, 2)$$

for the two characters.

Under the RR scheme outlined above, each person chosen from the hth stratum reports his or her $y^{(j)}$ value [minus a constant $\xi_h^{(j)}$] with probability C or one of the numbers $X_{h1}^{(j)}, \ldots, X_{h\mathrm{M}_{jh}}^{(j)}$ each with probability $M_{jh}^{-1}(1-C)$ $(j=1, 2)$, where

$$\sum_{u=1}^{\mathrm{M}_{jh}} X_{hu}^{(j)}=0$$

for each j and h. Both the quantities $\xi_h^{(j)}$ and the sets $S_{jh}=\{X_{h1}^{(j)}, \ldots, X_{h\mathrm{M}_{jh}}^{(j)}\}$ of anticipated values are preassigned. Recalling that in this particular survey, the box of beads used for randomization contained 44 red and 36 black beads with a person reporting his or her true value if and only if a red bead was observed, it is clear that $C=44/80=0.55$.

Thus for $j=1, 2$; $h=1, 2, \ldots, 5$, if the ith individual in the hth stratum is chosen in the sample, his or her response relating to the jth character may be denoted by $Z_{hi}^{(j)}$, where

$$\begin{aligned} Z_{hi}^{(j)} &= Y_{hi}^{(j)} && \text{with probability } C \\ &= X_{hu}^{(j)} && \text{with probability } M_{jh}^{-1}(1-C) \qquad (u=1, \ldots, M_{jh}) \end{aligned}$$

Let s_h be the sample from the hth stratum and π_{hi} the inclusion probability of the ith ($i=1, \ldots, N_h$) individual in this stratum ($h=1, \ldots, 5$). Under the

IPPS sampling scheme used, $\pi_{hi}>0$ for each i and h. Then suitably modifying the Horvitz-Thompson estimator in the present RR setup, for $j=1, 2; h=1,\ldots, 5$, a UE of $Y_h^{(j)}$, based on the sample data $(s_h, Z_{hi}^{(j)} \mid i \in s_h)$, is given by

$$\hat{Y}_h^{(j)} = \frac{1}{C} \sum_{i \in s_h} \frac{Z_{hi}^{(j)}}{\pi_{hi}} + \xi_h^{(j)}$$

It has a variance

$$\begin{aligned}\operatorname{var}(\hat{Y}_h^{(j)}) &= \sum_1^{N_h} (Y_{hi}^{(j)})^2 \left(\frac{1}{C\pi_{hi}} - 1\right) \\ &+ \sum_{i \neq t}^{N_h}\sum \left(\frac{\pi_{hit}}{\pi_{hi}\pi_{ht} - 1}\right) Y_{hi}^{(j)} Y_{ht}^{(j)} \\ &+ \frac{1-C}{C^2}\left[\frac{1}{M_{jh}} \sum_{u=1}^{M_{jh}} (X_{hu}^{(j)})^2\right] \sum_1^{N_h} \frac{1}{\pi_{hi}}\end{aligned}$$

with π_{hit} as the joint inclusion probability of the ith and tth individuals in the hth stratum. A UE of the variance above is provided by

$$\begin{aligned}\hat{\operatorname{var}}(\hat{Y}_h^{(j)}) &= \left(\frac{1}{C^2} \sum_{i<t \in s_h}\sum\right) \left(\frac{\pi_{hi}\pi_{ht}}{\pi_{hit}} - 1\right) \left(\frac{Z_{hi}^{(j)}}{\pi_{hi}} - \frac{Z_{ht}^{(j)}}{\pi_{ht}}\right)^2 \\ &+ \frac{1-C}{C^2} \sum_{i \in s_h} \frac{(Z_{hi}^{(j)})^2}{\pi_{hi}} + \frac{1-C}{C} \frac{N_h}{M_{jh}} \sum_{u=1}^{M_{jh}} (X_{hu}^{(j)})^2\end{aligned}$$

provided that $\pi_{hit}>0$ for each h, i, *and* t. Finally, for $j=1, 2$, a UE of the

TABLE 9.2. Strata sizes and corresponding sample sizes

Serial Number of stratum h	Stratum size N_h	Sample size n_h
1	20	10
2	22	11
3	10	5
4	18	9
5	14	7

TABLE 9.3. Certain design parameters for various strata

h	j	$\xi_h^{(j)}$ (rupees)	$S_{jh}=\{X_{h1}^{(j)}\ldots, X_{M_{jh}}^{(j)}\}$ (rupees)
1	1	300	$0, \pm 10, \pm 20, \ldots, \pm 200, \pm 300$
1	2	150	$0, \pm 10, \pm 20, \ldots, \pm 100, \pm 150$
2	1	550	$0, \pm 10, \pm 20, \ldots, \pm 450, \pm 550$
2	2	350	$0, \pm 10, \pm 20, \ldots, \pm 300, \pm 350$
3	1	300	$0, \pm 10, \pm 20, \ldots, \pm 200, \pm 300$
3	2	150	$0, \pm 10, \pm 20, \ldots, \pm 100, \pm 150$
4	1	300	$0, \pm 10, \pm 20, \ldots, \pm 50, \pm 300$
4	2	300	$0, \pm 10, \pm 20, \ldots, \pm 100, \pm 300$
5	1	350	$0, \pm 10, \pm 20, \ldots, \pm 250, \pm 350$
5	2	300	$0, \pm 10, \pm 20, \ldots, \pm 200, \pm 300$

TABLE 9.4. Stratum-wise and pooled estimates (in rupees), and variance estimates

Serial Number of Stratum	$\hat{Y}_h^{(1)}$	$\hat{v}\text{ar}(\hat{Y}_h^{(1)})$	$\hat{Y}_h^{(2)}$	$\hat{v}\text{ar}(\hat{Y}_h^{(2)})$
1	7,832.91	3.85×10^7	1,117.73	0.13×10^7
2	13,773.87	23.22×10^7	−706.05[a]	0.64×10^7
3	14,644.22	7.30×10^7	3,157.04	1.49×10^7
4	21,795.40	36.32×10^7	4,614.24	0.45×10^7
5	2,577.07	0.17×10^7	192.27	0.36×10^7
Population	60,623.47	70.86×10^7	8,375.23	3.07×10^7

[a] An interpretation for this negative value is that the corresponding anticipated values were possibly not chosen appropriately.

population total of $y^{(j)}$ is given by $\hat{Y}^{(j)}=\sum_h \hat{Y}_h^{(j)}$ and its variance is unbiasedly estimated by

$$\hat{v}\text{ar}(\hat{Y}^{(j)}) = \textstyle\sum_h \hat{v}\text{ar}(\hat{Y}_h^{(j)})$$

As indicated earlier, the gross salary in June 1982 was the basis of stratification. The salary ranges for strata 1, 2, 3, 4, and 5 were 500 to 899, 900 to 1599, 1600 to 2099, 2100 to 2599, and 2600 to 4100 rupees, respectively. From each stratum, sampling was done following an IPPS scheme. The details of the inclusion probabilities, which were obtained taking the net take-home salaries of June 1982 as size measures, are somewhat uninteresting and are not presented here. Table 9.2 presents the stratum sizes and the sample sizes for the various strata. Certain relevant design parameters for the strata are shown in Table 9.3. The survey results are summarized in Table 9.4, which shows the unbiased estimates of the stratum totals and the grand totals of the two sensitive characters studied. Unbiased estimates of the variances are also indicated.

9.3. RANDOMIZED RESPONSE TECHNIQUE TO DETERMINE INPUT IN CROP ESTIMATION

Chaudhuri (1983) has reported an application of the RR technique to estimate farm inputs needed for the development of a production function model connecting farm input with crop output. For the construction of such a model, one requires knowledge of the error distributions of both input items and outputs. In practice it is not difficult to obtain reliable data on the output of a crop by field inquiries. Comparing the estimated true output with what a farmer reports, one may fit an appropriate output error distribution. But administering inputs such as fertilizer, irrigation, and labor is a long-drawn-out process and one has hardly any way to check the validity of reported input figures, which are generally drastically over-reported. So an RRT may be applicable in the following way, suggested by Chaudhuri (1983). The population of farmers may be grouped into as large a number of strata as possible, with internal homogeneity in terms of amount of land cultivated, type of land, economic condition of the cultivator, amounts of manure/fertilizer used, extent of irrigation, amounts of expenditures on other efforts in tilling the soil, and so on. He suggested applying the modified Horvitz-Thompson method (see Section 7.3) of estimating the strata means for each item of input, with data procured thereon by RR techniques following methods described in Chapter 7. He then suggested taking each estimated stratum mean as the true input value for each member of the corresponding stratum. Comparing the reported values of input items through direct surveys with the estimated values, rather than imputed values, one may suppose to have uncovered the errors and proceed to study their distributions and follow up to postulate the production function models. Details are omitted.

REFERENCES

Adhikary, A. K. (1982) on RR surveys with sensitive quantitative characters: a case study in Indian Statistical Institute. *Tech. Rep. ASC*/82/6, Indian Statistical Institute, Calcutta.

Chaudhuri, A. (1983). RR technique to determine input in crop estimation. In: Proceedings of the Symposium on Statistical Methodology for Crop Forecasting and Crop Estimation. *Bull. Calcutta Statist. Assoc.* **32**, 208–210.

Mukhopadhyay, P. (1980). A survey on the socio-economic conditions of some college students of Calcutta (with special emphasis on drug habits of students). Project report, Indian Statistical Institute, Calcutta.

Mukhopadhyay, P., and Halder, A. K. (1980). Bayesian tables for Warner's RR probabilities. *Tech. Rep. ASC*/80/2, Indian Statistical Institute, Calcutta.

Sengupta, S. (1981). A sampling scheme to realize inclusion probabilities exactly proportional to size. *J. Indian Soc. Agri. Statist.* **33**, 19–23.

Appendix 4

Overview of Unified Theory of Direct Surveys

A4.1 INTRODUCTION AND NOTATION

We present here a brief résumé of some salient features of the unified theory of varying probability sampling in direct surveys of labeled finite populations with an emphasis on Horvitz-Thompson (1952) estimation as it affects the development of RR strategies. The aim is to provide the reader with summarized materials culled from the literature on open surveys so as to facilitate an appreciation of the corresponding developments relating to RR surveys that we reported in Chapter 7.

Let $\underline{Y} = (Y_1, \ldots, Y_i, \ldots, Y_N)$ be the vector of values Y_i on a real variate y defined on a labeled finite population $I_N = (1, \ldots, i, \ldots, N)$ of a given number N of individuals. Let s be a sample which is a sequence of labels $(i_1, \ldots, i_n)$, not necessarily distinct, arranged in order of draws from I_N. Let this be drawn with probability $P(s)$. The probability measure P will be called a (sampling) design. If every s has a fixed number, n (say), of distinct labels, we write P_n for P to denote a fixed (effective) sample size (n) design. When s is chosen and the Y_i values, $i \in s$, are observed, the sequence $d = ((i_1, Y_{i_1}), \ldots, (i_n, Y_{i_n}))$ constitutes the survey data at hand.

A real-valued function $t = t(d) = t(s, \underline{Y})$ involving Y_i only for $i \in s$ but not $i \notin s$ is to be used as an estimator for $Y = \Sigma Y_i$ together with any other relevant and available data: for example, values on auxiliary variables x, z, and so on, related to y in appropriate manners. This t is (design or P-) unbiased for Y if

$$E(t) = E_p(t) = \sum_s t(d)P(s) = Y$$

for every $\underline{Y}$. Here E_p denotes the operator for taking expectation with respect to P. Similarly, we will write V_p and C_p to denote operators for taking variance and covariance with respect to the design P.

A linear unbiased estimator (LUE) for Y is of the form

$$t = t(d) = a_s + \sum_{i \in s} b_{si} Y_i = t_L \quad \text{(say)}$$

such that $b_{si} = 0$ if $i \notin s$,

$$\sum a_s P(s) = 0 \qquad \sum_{s \ni i} b_{si} P(s) = 1 \qquad (i = 1, \ldots, N)$$

and a_s and b_{si} free of $\underline{Y}$. Here $\Sigma_{s \ni i}$ denotes the sum over samples containing the label i.

In case $a_s = 0$, we call t a homogeneous linear unbiased estimator (HLUE) for Y and denote it by t_H. In particular,

$$\bar{t} = \sum_{i \in s} \frac{Y_i}{\pi_i}$$

is the Horvitz-Thompson estimator (HTE), denoting by $\pi_i = \Sigma_{s \ni i} P(s)$ the inclusion probability of i in a sample chosen according to P and assuming (throughout) that $\pi_i > 0, i = 1, \ldots, N$.

A design P is called a unicluster design (UCD) if any two samples s_1 and s_2 with $P(s_1) > 0$, $P(s_2) > 0$, are either disjoint or equivalent, containing a common set of distinct units with no uncommon units. Any other design is a non-unicluster design (NUCD). A combination of a design P and an estimator t based on it will be called a (sampling) strategy (P, t).

A4.2 ASSORTMENT OF LEADING THEORETICAL RESULTS

Godambe (1955) simplified the problem of sampling finite populations by showing that for a design P, in general, in the class of HLUE for Y, one does not exist with the least variance irrespective of the variate values.

Godambe (1965), Hege (1965), Hanurav (1966), and Lanke (1975) observed that a necessary and sufficient condition for a uniformly minimum variance (UMV) estimator for Y to exist in the HLUE class is that the design be a UCD. For such a design, the UMV estimator is the HTE.

Godambe and Joshi (1965) showed that in the class of all unbiased estimators (UE) for Y, a UMV estimator does not exist for any NUCD. Basu (1971) extended this nonexistence result to every noncensus design, that is, any design that does not assign the entire selection probability to the whole population I_N itself. Basu's (1971) result does not imply Godambe's (1955), or vice versa.

In the class of HLUE for Y the HTE is admissible; that is, for no other estimator t_H for Y do the variances satisfy

$$V_p(t_H) \leq V_p(\bar{t}) \qquad \text{for every } \underline{Y}$$

and strictly so for at least one $\underline{Y}$. This result was proved independently by Godambe (1960) and by Roy and Chakravarti (1960), and extended by Godambe and Joshi (1965) to cover the entire UE class.

An effective way to find optimal estimators and/or strategies is to adopt a superpopulation approach. A superpopulation concept consists in regarding $\underline{Y}$ as a random vector with a certain probability distribution. This probability distribution is supposed to lay down a superpopulation model. The expectation, variance, and covariance operators for such modeling are denoted by E_m, V_m, and C_m respectively. In this approach, performances of estimators or strategies are assessed in terms of the criterion $E_m V_p(t)$. It is customary to demand that $E_p(t) = Y$ in this case also. But sometimes one imposes, instead, the requirement that $E_m\,(t - Y) = 0$, for every s with $P(s) > 0$, called model (or m) unbiasedness. More generally, one is inclined to demand that $E_m E_p\,(t - Y) = 0$, called model-design (mP-) unbiasedness. We will assume that E_m and E_p commute.

Let us write M to denote a general model and M_j $(j = 1, 2, \ldots)$ its special cases illustrated below together with associated results prominent in the literature.

Model M: $E_m(Y_i) = \mu_i$, $V_m(Y_i) = \sigma_i^2$, $C_m(Y_i, Y_j) = \rho\sigma_i\sigma_j\,(i, j = 1, \ldots, N;\ i \neq j)$. The parameters $\mu_i\,(-\infty < \mu_i < \infty)$, $\sigma_i^2\,(>0)$, and $\rho[-1/(N-1) < \rho < 1]$ are otherwise unknown.

Model M_1: $\mu_i = \alpha_i + \beta X_i$, $\sigma_i^2 = \sigma^2 X_i^2$ with α_i, X_i (>0) known but β, $\sigma^2(>0)$ unknown.

Model M_2: $C_m(Y_i, Y_j) = 0$ $(i \neq j)$, μ_i of a common sign.

Model M_3: Y_i, $i = 1, \ldots, N$, are independently distributed.

Model M_4: A special case of M_3 with

$$\mu_i = \alpha_i + \beta X_i \qquad \sigma_i^2 = \sigma^2 f_i \qquad \sigma > 0 \qquad f_i > 0 \qquad X_i > 0$$

and α_i, X_i, f_i, and β known, but σ unknown.

Model M_5: Same as M_2 with $\mu_i = \beta X_i$, $\sigma_i^2 = \sigma^2 X_i^g$, β, σ, $X_i > 0$, and $g \in [0, 2]$.

Any P_n design with $\pi_i \propto \mu_i$ will be denoted as $P_{n\mu}$, any design P_n with $\pi_i \propto X_i$ will be written P_{nX}, and any design $P_{n\mu}$ with $\pi_i \propto \mu_i \propto \sigma_i$ will be denoted by $P_{n\mu\sigma}$. For a P_{nX} we will write $\pi_i = \pi_i(X)$. The following results are considered worthy of reference. Under M_1, for any LUE t_L and P_{nX},

$$E_m V_{p_n}(t_L) \geq (1-\rho)(1-fA)\frac{\sigma^2}{n} = E_m V_{p_{nX}}(\bar{\bar{t}}\,(X))$$

writing $A = \Sigma X_i^2/N$, $X = \Sigma X_i = N$, $f = n/N$, $\alpha = \Sigma \alpha_i$, and

$$\bar{\bar{t}} = \bar{\bar{t}}(X) = \sum\nolimits_{i \in s} \frac{Y_i - \alpha_i}{\pi_i(X)} + \alpha$$

This result is due to Cassel *et al.* (1976). An estimator of the form $\bar{\bar{t}} = \Sigma_{i \in s}(Y_i - \alpha_i)/\pi_i + \alpha$ for any P with $\pi_i > 0$, called a generalized difference estimator, was introduced by Basu (1971) and studied extensively by Cassel *et al.* (1977).

Under M_2, from Godambe (1955):

(a) $E_m V_{p_{n\mu}}(t_L) \geq E_m V_{p_{n\mu}}(\bar{t})$.
(b) $E_m V_{p_n}(t_L) \geq E_m V_{p_{n\mu\sigma}}(\bar{t})$.
Under M_3, writing $\mu = \Sigma \mu_i$ and $t^* = \Sigma_{i \in s}(Y_i - \mu_i)/\pi_i + \mu$:

1. $E_m V_p(t) \geq E_m V_p(t^*)$, due to Ho (1980).
2. $E_m V_p(t) \geq \Sigma \sigma_i^2\ (1/\pi_i - 1)$, $E_m V_{p_{n\mu}}(t) \geq E_m V_{p_{n\mu}}(\bar{t})$, and $E_m V_{p_n}(t) \geqslant E_m V_{p_{n\mu\sigma}}(\bar{t}) = (\Sigma\sigma_i)^2/n - \Sigma_i \sigma_i^2 = \xi$ (say), due to Godambe and Joshi (1965).
3. A necessary and sufficient condition for t to attain the lower bound ξ is that t is t^* based on a P_n design.
4. A necessary and sufficient condition for $\bar{t}$ to attain the lower bound ξ is that it is m-unbiased and is based on a P_n design; results (3) and (4) are due to Ho (1980).

Under M_4, a necessary and sufficient condition that $E_m V_p(t)$ for any P_n and UE t for Y be the minimum is that $(P, t) = (P', t')$, where P' is a P_n design with

$$\pi_i = \frac{nf_i}{\sum_1^N f_i}$$

and

$$t' = \sum\nolimits_{i \in s} \frac{Y_i - \alpha_i - \beta X_i}{\pi_i} + \sum_1^N (\alpha_i + \beta X_i)$$

If certain details about $((i, Y_i)|i \in s)$ are unavailable at analysis or are discarded, but Y_i/π_i, $i \in s$, are available, then Särndal (1972, 1976) shows that among all functions of $(Y_i/\pi_i,\ i \in s)$, the HTE is unique among P-unbiased estimators for Y and hence the UMV among them.

Certain optimality results are also available on postulating random permutation models $\mathscr{P}$, and exchangeable models $\mathscr{E}$, on which important reported works are due to Kempthorne (1969), Rao (1971), and Godambe and Thompson (1973), among others. Model $\mathscr{E}$ is as follows:

Model $\mathscr{E}$: For a fixed vector $\underline{A}=(A_1, \ldots, A_i, \ldots, A_N)$, with $0<A_i<1$, $i=1, \ldots, N$, $\Sigma A_i=n$ (fixed sample size) and $r_i=Y_i/A_i$, $i=1, \ldots, N$, the vector $\underline{r}=(r_1, \ldots, r_i, \ldots, r_N)$ has an exchangeable distribution.

A P_n design with $\pi_i=A_i$ will be denoted as P_{nA}. The following results are due to Godambe and Thompson (1973). Under model $\mathscr{E}$:

1. For every P_{nA}, the value of $E_m V_{P_{nA}}(\bar{t})$ is common.
2. $E_m V_{P_n}(t) \geq E_m V_{P_{nA}}(\bar{t})$.

A special case of $\mathscr{E}$ is the permutation model $\mathscr{P}$, stipulating $\underline{r}$ as a random realization out of $N!$ vectors realizable on permuting the coordinates of $\underline{r}$. This model was introduced by Kempthorne (1969) for obtaining optimality results in finite population inference. Rao (1971) utilized it in obtaining an optimal property for the HTE based on any P_n under a model analogous to the one above due to Godambe and Thompson (1973) under model $\mathscr{E}$.

Rao and Bellhouse (1978) considered a generalization of the model $\mathscr{E}$ permitting observational errors in Y_i, $i=1, \ldots, N$. Their optimality result is analogous to the one above due to Godambe and Thompson, with the modification that $\bar{t}$ takes the form

$$\bar{t}' = \sum_{i \in s} \frac{Y_i'}{\pi_i}$$

where Y_i' is Y_i subject to errors of observations such that the expected value of Y_i' calculated with the probability distribution of the variable denoting observational errors equals Y_i, $i=1, \ldots, N$.

In addition to these optimality results, there are certain others concerning relative performances of specific sampling strategies and methods of estimating variances of estimators for different strategies. Let us mention a few of them for which we reported corresponding parallel theory applicable to RR situations discussed in Chapter 7.

The sampling scheme due to Rao *et al.* (1962) consists of (1) forming in groups taking N_i labels by the SRSWOR method from I_N, $i=1, \ldots, n$, such that $\Sigma_{i=1}^n N_i=N$, and (2) choosing just one label i_j $(j=1, \ldots, N_i)$, falling in the ith group with a probability P_{i_j}/θ_i, independently from each of n groups so formed. We write P_i as the known positive normed size measure for a label i in I_N and $\theta_i = \Sigma_{j=1}^{N_i} P_{i_j}$. The estimator is

$$t_{RHC} = \sum_{i=1}^{n} \frac{Y_{i_j}}{P_{i_j}} \theta_i$$

We will assume for simplicity that N/n is an integer and that $N_i = N/n$ for every $i = 1, \ldots, n$.

Lahiri-Midzuno-Sen (1951, 1952, 1953) strategy consists of using the ratio estimator

$$t_R = \frac{\sum_{i \in s} Y_i}{\sum_{i \in s} P_i}$$

for Y based on the selection scheme which chooses on the first draw a label i with probability P_i following it up with an SRSWOR of $(n-1)$ units from the other $(N-1)$ labels of I_N, excluding i.

The Hansen-Hurwitz (1943) strategy chooses units with replacement in n draws with a selection probability P_i for label i on each draw and employs for Y the estimator

$$t_{HH} = \sum_{r=1}^{n} \frac{y_r}{np_r}$$

denoting by $y_r(p_r)$ the $Y_i(P_i)$ value for the label chosen on the rth draw.

The symmetrized Des Raj (1956) strategy due to Murthy (1957) chooses two labels from I_N with a probability P_i for i taken on the first draw and a probability $P_j/(1-P_i)$ for a label $j \, (\neq i)$ on the second draw given that the first draw yielded the label i. The unbiased estimator for Y based on such a sample is

$$t_M = \frac{Y_i(1-P_j)/P_i + Y_j(1-P_i)/P_j}{2-P_i-P_j}$$

Also note the following:

$$V_p(\bar{t}) = \sum \frac{Y_i^2}{\pi_i} + \sum_{i \neq j}\sum Y_i Y_j \frac{\pi_{ij}}{\pi_i \pi_j} - Y^2 \qquad \text{(for any } P \text{ with } \pi_{ij} > 0)$$

$$= \sum_{1<j}\sum \left(\frac{Y_i}{\pi_i} - \frac{Y_j}{\pi_j} \right)^2 (\pi_i \pi_j - \pi_{ij}) \qquad \text{(for any } P_n \text{ with } \pi_{ij} > 0)$$

writing $\pi_{ij} = \Sigma_{s \supset ij} P(s) \equiv$ the inclusion probability of the pair of distinct lables i and j.

Assuming that a fixed sample size (n, say) will be used for every strategy under consideration, a usual unbiased estimator for $V(\bar{t})$, due to Yates and Grundy (1953), is

$$v = \sum\sum_{i<j \in s} \frac{(Y_i/\pi_i - Y_j/\pi_j)^2 (\pi_i \pi_j - \pi_{ij})}{\pi_{ij}}$$

The following variance formulas are well known:

$$V_p(t_{\mathrm{RHC}}) = \frac{\sum N_i^2 - N}{N(N-1)} \sum\sum_{i<j} \left(\frac{Y_i}{P_i} - \frac{Y_j}{P_j} \right)^2 P_i P_j$$

$$V_p(t_{\mathrm{R}}) = \sum\sum_{i<j} \left(\frac{Y_i}{P_i} - \frac{Y_j}{P_j} \right)^2 P_i P_j \left(1 - \frac{\sum_{i \in s} P_i}{\binom{N-1}{n-1}} \right)$$

$$V_p(t_{\mathrm{HH}}) = \sum\sum_{i<j} \left(\frac{Y_i}{P_i} - \frac{Y_j}{P_j} \right)^2 \frac{P_i P_j}{n(n-1)}$$

$$V_p(t_{\mathrm{M}}) = \frac{1}{2} \sum\sum_{i<j} \left(\frac{Y_i}{P_i} - \frac{Y_j}{P_j} \right)^2 \frac{P_i P_j (1 - P_i - P_j)}{2 - P_i - P_j}$$

For unbiased variance estimation, the following result due to Rao (1979) is worthy of attention. If a homogeneous linear estimator for Y, namely, $t_{\mathrm{HL}} = \Sigma_{i \in s} b_{si} Y_i$, has a mean square error

$$M(t_{\mathrm{HL}}) = E_p (t_{\mathrm{HL}} - Y)^2$$

which equals zero if $Y_i \propto W_i \ (\neq 0)$, then one has

$$M(t_{\mathrm{HL}}) = -\sum\sum_{i<j} W_i W_j d_{ij} (Z_i - Z_j)^2$$

Here $Z_i = Y_i / W_i$ and $d_{ij} = E_p (b_{si} - 1)(b_{sj} - 1)$. A design-unbiased estimator for $M(t_{\mathrm{HL}})$ is then

$$m(s) = -\sum\sum_{i<j \in s} W_i W_j d_{ij}(s) (Z_i - Z_j)^2$$

with $d_{ij}(s)$ such that $E_p d_{ij}(s) = d_{ij}$.

An interesting result under model M_5, taking $\pi_i = nP_i$, is

$$E_m V_p(\bar{t}) \lesseqgtr E_m V_p(t_{\mathrm{RHC}}) \lesseqgtr E_m V_p(t_{\mathrm{R}}) \qquad \text{if} \quad g \gtreqless 1.$$

This result is due to Chaudhuri and Arnab (1979) but was partially proved earlier by Rao (1966), Rao (1967), and Hanurav (1967).

REFERENCES

Basu, D. (1971). An eassay on the logical foundations of survey sampling, Part 1. In: *Foundations of Statistical Inferences*, ed. V. P. Godambe, and D. A. Sprott. Holt, Rinehart and Winston, Toronto, pp. 203–242.

Cassel, C. M., Särndal, C. E., and Wretman, J. H. (1976). Some results on generalized difference estimation and generalized regression estimation for finite populations. *Biometrika* **63**, 615–620.

Cassel, C. M., Särndal, C. E., and Wretman, J. H. (1977). *Foundations of Inference in Survey Sampling*. John Wiley, New York.

Chaudhuri, A., and Arnab, R. (1979). On the relative efficiencies of sampling strategies under a super population model. *Sankhyā Ser. C* **41**, 40–43.

Godambe, V. P. (1955). A unified theory of sampling from finite populations. *J. Roy. Statist. Soc. Ser. B.* **17**, 269–278.

Godambe, V. P. (1960). An admissible estimate for any sampling design. *Sankhyā* **22**, 285–288.

Godambe, V. P. (1965). A review of the contributions towards a unified theory of sampling from finite populations. *Internat. Statist. Rev.* **33**, 242–258.

Godambe, V. P., and Joshi, V. M. (1965). Admissibility and Bayes estimation in sampling finite populations, *I. Ann. Math. Statist.* **36**, 1707–1722.

Godambe, V. P., and Thompson, M. E. (1973). Estimation in sampling theory with exchangeable prior distributions. *Ann. Statist.* **1**, 1212–1221.

Hansen, M. H., and Hurwitz, W. N. (1943). On the theory of sampling from finite populations. *Ann. Math. Statist.* **14**, 333–362.

Hanurav, T. V. (1966). Some aspects of unified sampling theory. *Sankhyā Ser. A* **28**, 175–204.

Hanurav, T. V. (1967). Optimum utilization of auxiliary information: πPS sampling of two units from a stratum. *J. Roy. Statist. Soc. Ser. B* **29**, 374–391; **31**, 192–194 (1969) with a correction note.

Hege, V. S. (1965). Sampling designs which admit uniformly minimum variance unbiased estimators. *Bull. Calcutta Statist. Assoc.* **14**, 160–162.

Ho, E. W. H. (1980). Model-unbiasedness and the Horvitz-Thompson estimator in finite population sampling. *Austral. J. Statist.* **22**, 218–225.

Horvitz, D. G., and Thompson, D. J. (1952). A generalization of sampling without replacement from a finite universe. *J. Amer. Statist. Assoc.* **47**, 663–685.

Kempthorne, O. (1969). Some remarks on statistical inference in finite sampling. In: *New Developments in Survey Sampling*, ed. N. L. Johnson, and H. Smith, Jr. Wiley-Intersciences, New York, pp. 671–675.

Lahiri, D. B. (1951). A method of sample selection providing unbiased ratio estimators. *Bull. Internat. Statist. Inst.* **33**, II, 133–140.

Lanke, J. (1975). Some contributions to the theory of survey sampling. Ph.D. thesis, University of Lund.

Midzuno, H. (1952). On the sampling system with probabilities proportional to sum of sizes. *Ann. Inst. Statist. Math.* **3**, 99–107.

Murthy, M. N. (1957). Ordered and unordered estimators in sampling without replacement. *Sankhyā* **18**, 379–390.

Raj, Des (1956). Some estimators in sampling with varying probabilities without replacement. *J. Amer. Statist. Assoc.* **51**, 269–284.

Rao, J. N. K. (1966). On the relative efficiency of some estimators in PPS sampling for multiple characters. *Sankhyā Ser. A* **28**, 61–70.

Rao, T. J. (1967). On the choice of a strategy for the ratio method of estimation. *J. Roy. Statist. Soc. Ser. B* **24**, 482–497.

Rao, C. R. (1971). Some aspects of statistical inference in problems of sampling from finite populations. In *Foundations of Statistical Inferences*, ed. V. P. Godambe and D. A. Sprott, Holt, Rinehart and Winston. Toronto, pp. 177–202.

Rao, J. N. K. (1979). On deriving mean square errors and their non-negative unbiased estimators in finite population sampling. *J. Indian Statist. Assoc.* **17**, 125–136.

Rao, J. N. K., and Bellhouse, D. R. (1978). Optimal estimation of a finite population mean under generalized random permutation model. *J. Statist. Plann. Inference* **2**, 125–147.

Rao, J. N. K., Hartley, H. O., and Cochran, W. G. (1962). On a simple procedure of unequal probability sampling without replacement. *J. Roy. Statist. Soc. Ser. B* **24**, 482–491.

Roy, J., and Chakravarti, I. M. (1960). Estimating the mean of a finite population. *Ann. Math. Statist.* **31**, 392–398.

Särndal, C. E. (1972). Sample survey theory vs. general statistical theory: estimation of the population mean. *Internat. Statist. Rev.* **40**, 1–12.

Särndal, C. E. (1976). On uniformly minimum variance estimation in finite populations. *Ann. Statist.* **4**, 993–997.

Sen, A. R. (1953). On the estimate of the variance in sampling with varying probabilities. *J. Indian. Soc. Agri. Statist.* **5**, 119–127.

Yates, F., and Grundy, P. M. (1953). Selection without replacement from within strata with probability proportional to size. *J. Roy. Statist. Soc. Ser. B* **15**, 253–261.

Index